Biostatistics and Computer-based
Analysis of Health Data using R

Biostatistics and Health Science Set

coordinated by
Mounir Mesbah

Biostatistics and Computer-based Analysis of Health Data using R

Christophe Lalanne
Mounir Mesbah

ELSEVIER

First published 2016 in Great Britain and the United States by ISTE Press Ltd and Elsevier Ltd

ISTE Press Ltd
27-37 St George's Road
London SW19 4EU
UK

www.iste.co.uk

Elsevier Ltd
The Boulevard, Langford Lane
Kidlington, Oxford, OX5 1GB
UK

www.elsevier.com

For information on all our publications visit our website at http://store.elsevier.com/

British Library Cataloguing-in-Publication Data
A CIP record for this book is available from the British Library
Library of Congress Cataloging in Publication Data
A catalog record for this book is available from the Library of Congress
ISBN 978-1-78548-088-1

Printed and bound in the UK and US

Contents

Introduction

A large number of the actions performed by means of statistical software amount to manipulating or even to transforming digital data representing statistical data literally. It is therefore paramount we understand how statistical data are represented and how they can be used by software such as R. After importing, recoding and the eventual transformation of these data, the description of the variables of interest and the summary of their distribution in numerical and graphical form constitute a prior and fundamental step to any statistical modeling, hence the importance of these early stages in a statistical analysis project. In the second step, it is essential to fully control the commands that enable the calculation of the main measures of association in medical research and to know how to implement the conventional explanatory and predictive models: variance analysis, linear and logistic regression and Cox model. With few exceptions, using the R commands available during the installation of the software (base commands) is favored over the use of specialized commands in external R packages. The packages that must be installed to follow the applications presented in this book are listed in Chapter 1, in section 1.1.

This book assumes that the reader is already familiar with basic statistical concepts, particularly the calculation of central tendency and dispersion indicators for a continuous variable, contingency tables, analysis of variance and conventional regression models. The objective is to apply this knowledge using data sets described in numerous other works, even if the interpretation of the results remains minimal, and familiarize oneself quickly with the use of R with actual data. Emphasis is given to the management and manipulation of structured data, as this constitutes 60 to 80% of the work of the statistician. There are many works in French and in English on R, both from the technical and statistical point of view. Some of these works are oriented towards general aspects [SHA 12], others are much more specialized [BIL 14] or address more advanced concepts [HOT 09]. The purpose of this book is to enable the reader to get accustomed to R so that he can perform his own analyses and continue his apprenticeship in an autonomous way in the field of medical statistics.

In Chapter 1, the base commands for the management of data with R are introduced. The focus is on the creation and manipulation of quantitative and qualitative variables (recoding individual frequencies, counting missing observations), importing databases stored in the form of text files, and basic arithmetic operations (minimum, maximum, arithmetic mean, difference, frequency, etc.). We will also consider how to store preprocessed databases in text or in R formats. The objective is to understand how the data are represented in R and how to work with them.

Chapter 2 focuses on useful commands for the description of a data table comprised of quantitative or qualitative variables. The descriptive approach is strictly univariate, which is the prerequisite for any statistical approach. Basic graphic commands (histograms, density curves, bars or points diagrams) will be presented in addition to the usual central tendency (mean, median) and dispersion (variance, quartiles) descriptive summaries. Pointwise and interval estimation using means and empirical proportions will also be addressed. The objective is to become acclimatized with the use of simple R commands operating on a variable, optionally specifying some options for the calculation, and also with the selection of statistical units.

Chapter 3 is devoted to the comparison of two samples for quantitative or qualitative measurements. The following hypothesis tests are addressed: Student's t-test for independent or paired samples, non-parametric Wilcoxon test, χ^2 test and Fisher's exact test, McNemar test, from the main association measurements for two variables (mean difference, odds ratio and relative risk). From this chapter on, there will be less emphasis on the univariate description of each variable, but it is advisable to carry out exploratory data analysis as discussed in Chapter 2. The objective is to control the main statistical tests in the case where we are interested in the relationship between a quantitative variable and a qualitative variable or in the case of two qualitative variables.

Chapter 4 is an introduction to the analysis of variance (ANOVA) in which the objective is to explain the variability observed at the level of a numeric response variable by taking into account a group or a classification factor and mean differences interval estimation. We will focus on the construction of an ANOVA table summarizing the various sources of variability and on the graphic methods allowing us to summarize the distribution of individual or aggregated data. The linear tendency test will also be studied in the case where the classification factor can be considered as being naturally ordered. The objective is to understand how to construct an explanatory model where there is one or even two explanatory factors and present the results of such a model through the use of R digitally and graphically.

Chapter 5 focuses on the analysis of the linear relationship between two continuous quantitative variables. In the linear correlation approach, which assumes a symmetrical relation between the two variables, we are interested in the quantification of the magnitude and direction of the association in a parametric (Pearson correlation) or non-parametric manner (rank-based Spearman correlation) and on the graphic representation of this relation. Simple linear regression will be used in the event that one of the two numeric variables assumes the function of a response variable and the other one of an explanatory variable. The useful commands for the estimation of the coefficients of the regression line, the construction of the ANOVA table associated with the regression and the prediction will be presented. The objective of this chapter remains identical to that of Chapter 4, i.e. to present the R commands necessary for the construction of a simple statistical model with two variables following an explanatory or predictive perspective.

In Chapter 6, the main measures of association found in epidemiological studies will be discussed: odds ratio, relative risk, prevalence, etc. R commands allowing the (pointwise and interval-based) estimation and the associated hypothesis tests will be illustrated with cohort or case-control study data. The implementation of a simple logistic regression model makes it possible to complete the range of statistical methods that allow the variability observed to be explained at the level of binary response variables. The aim is to understand which R commands to use when the variables are binary, to summarize a contingency table in the form of association indicators or to model the relationship between a binary response (poor/healthy) and a qualitative explanatory variable from so-called grouped data.

The final chapter provides an introduction to the analysis of censored data, to the main tests related to the construction of a survival curve (log-rank or Wilcoxon tests) and finally to the Cox regression model. The specificity of censored data requires particular care in the coding of data in R and the objective is to present the R commands essential to the correct representation of survival data in digital form, to their digital (median survival) and graphical (Kaplan–Meier curve) summary, and the implementation of common tests of risk measures.

At the end of each chapter, a few applications are provided with a few examples of commands that can be used to respond to most questions. It is sometimes possible to obtain identical results by using other approaches or other commands. Outputs from R are not reproduced but the reader is encouraged to try the proposed R instructions and alternative or additional instructions. It will be assumed that the data files being used are available in the working directory. All of the data files and the R commands used in this book can be downloaded from https://github.com/biostatsante.

The three appendices will help familiarize the reader with RStudio, lattice packages for the management of graphical outputs and Hmisc and rms for advanced data management and modeling. These appendices are obviously not a substitute for the work of John Verzani, Paul Murrell and Franck Harrell [VER 11, MUR 05, HAR 01].

Due to the design of the layout, some R outputs have been truncated or reformatted. Therefore, there could be differences when the reader attempts to reproduce the commands of this book.

An index of the R commands used in the illustrations is available at the end of the book.

Elements of the Language

R is more than a simple software program for statistics; it is a language for the manipulation of statistical data [IHA 96, VEN 02]. This partly explains its difficult non user-friendly approach for users accustomed to drop-down menus such as those offered by SPSS (although SPSS also offers a basic macro language). This chapter allows the reader to discover the elements of the language and to become familiarized with the mechanisms by which to represent statistical data in R. In the illustrations that follow, the R commands are prefixed with the symbol >, which designates the R console prompt. It is, therefore, unnecessary to copy this symbol to test the proposed instructions.

1.1. Before proceeding

1.1.1. *Installing R*

The installation of R is relatively easy and instructions can be found at the following website: http://cran.r-project.org. The software program is available for Windows, Linux and Mac. In addition to the R program, the installer provides an R script editor, online help and a set of base packages. The packages include commands specific to a certain area (graphics commands, modeling commands, etc.) and make it possible to enhance the base features of R.

1.1.2. *RStudio*

Although R is sufficient to start or perform statistical analyses, RStudio (www. rstudio.com) provides a particularly enjoyable working environment for R. It includes a powerful R script editor, a console in which the user can execute the commands (or send those inserted in the script editor), online help, a graphics browser, a data viewer and much more [VER 11]. A brief introduction to the software is provided in the appendices.

1.1.3. *List of useful packages*

Although the R base packages allow for most statistical analyses addressed in this book, a number of additional packages must be installed to reproduce some of the proposed applications. In addition to the base packages that are used throughout this book (lattice, foreign, survival), etc. the following packages can be installed by means of the install.packages() command, in the format install.packages("reshape2"): reshape2, gridExtra, vcd (Chapter 3); epiR, car (Chapter 4); ppcor, psych (Chapter 5); epicalc, ROCR (Chapter 6).

These packages can be installed directly from RStudio using the package manager. Before installing a package, it is possible to consult online help on the CRAN website (http://cran.r-project.org). The CRAN website also offers the ability to search packages by theme ("Task Views").

To access the commands of a package, the command library() can be used, indicating the name of the package.

1.1.4. *Find help*

As with any piece of software, it is important to know how to find help on a specific command or from a keyword. The commands help() and help.search() or apropos() achieve these two functions in R. RStudio also provides an integrated search engine that facilitates access to the online help pages to a great extent.

1.1.5. *R scripts*

The easiest and recommended way to take advantage of the capabilities of RStudio in the management of a statistical analysis project is to enter the commands in the script editor while ensuring that these commands are saved in a R script file. The commands entered in the RStudio editor can be sent directly to the console and R will display the result in the same console. This interactive approach allows a command script to be incrementally built and enables the correction or adjustment of the controls as and when needed.

It is also important to document the main stages of the statistical analysis as well as commands that may not be clear to an external reader. In fact, one should always anticipate that the R command script could be reused in the future by a third person. Comments are useful in this case. You can add the prefix # to a line of text so that can it be treated as a comment and not as a statement by R. The commands in the R script should also be organized logically, effectively distinguishing the commands related to the importation of data, to univariate and bivariate descriptive statistics and to statistical models, etc. All this must be presented in very distinct sections and rely as little as possible on variables or tables of temporary data that would significantly alter the original data table, without any documentation or any storage to disk. The original data, as a matter of fact, should always be accessible at any point of the analysis. Intermediate data tables can be saved separately. In the case of large analysis projects, it is preferable to create several script files to store the commands corresponding to the different stages of the analysis.

1.2. Data representation in R

1.2.1. *Management of numerical variables*

Suppose that we have the weight of ten newborns available (variable x, in grams) and that of their mother (variable y, in kg), as illustrated in Table 1.1.

x	2523	2551	2557	2594	2600	2622	2637	2637	2663	2665
y	82.7	70.5	47.7	49.1	48.6	56.4	53.6	46.8	55.9	51.4

Table 1.1. *Artificial data about weights at birth*

In the following example, a variable called x was created in which the observations displayed in Table 1.1 are stored in the form of a simple list of numbers:

```
> x <- c(2523,2551,2557,2594,2600,2622,2637,2637,2663,2665)
```

The symbol <- is being used (in preference to =) to associate a series of measurements to a defined variable. These values can now be displayed in R using the print(x) command or by typing the name of the variable:

```
> x
[1] 2523 2551 2557 2594 2600 2622 2637 2637 2663 2665
```

We will make use of the same procedure with y:

```
> y <- c(82.7,70.5,47.7,49.1,48.6,56.4,53.6,46.8,55.9,51.4)
```

Spaces have no importance when a list of numbers and, with rare exceptions, other expressions are entered. The decimal separator is compulsorily the period, since R follows the English notation for numbers.

It can be verified that R has retained two variables in what is called the workspace with the help of the command ls():

```
> ls()
[1] "x"  "y"
```

Therefore, it is always possible to access these variables as long as the R session is not closed and as long as neither of these variables has been removed in the meantime.

The number of observations for variable x corresponds to the length of x (number of elements), when there are no missing values:

```
> length(x)
[1] 10
```

The values of a variable, or in more statistical terms, the values taken by a variable for a sample of size n, can be selected by specifying the index or number of the observation between square brackets. Section 1.2.2 will cover how it is possible to select more than one observation at a time and how observations can be selected on the basis of an external criterion and how this approach can be generalized to the selection of a set of observations gathered for the same statistical unit. The fifth observation is obtained in the following manner:

```
> x[5]
[1] 2600
```

The observations are indexed from 1 to n, where n is the total number of items in the variable under consideration. The first and the last observations will thus be obtained using the commands x[1] and x[10].

1.2.2. *Operations with a numerical variable*

It should be noted that the weights of the babies are expressed in grams, while those of the mother are in kg. The weight of the 5th baby in kg is obtained using a simple arithmetic division operation:

```
> x[5] / 1000
[1] 2.6
```

The same division can be applied to all of the observations, without having to carry division out for each element of x, as shown in the following example:

```
> x / 1000
[1] 2.523 2.551 2.557 2.594 2.600 2.622 2.637 2.637 2.663 2.665
```

The previous operation did not change the values of the original variable x, but it is easy to create a new variable in which the weight of the babies, expressed in kg, will be stored.

```
> x2 <- x / 1000
> x
[1] 2523 2551 2557 2594 2600 2622 2637 2637 2663 2665
> x2
[1] 2.523 2.551 2.557 2.594 2.600 2.622 2.637 2.637 2.663 2.665
```

It might also be desirable to replace the old values of x with the new values. In this case, we can replace the statement x2 <- x / 1000 with x <- x / 1000. However, it will not be possible to return to the old values of x after this operation is carried out. As when deleting a variable with the command rm(), any update of the elements of a variable is final for the session in progress.

In addition to the basic arithmetic operators (addition, subtraction, multiplication and division), R offers most of the functions found in a calculator: logarithm, exponential, absolute value, etc. Therefore, log(x) would return the values of x after transformation by the natural logarithm, whereas for the base 10 logarithm, log10(x) should be used.

Finally, R offers a wide range of commands that are more specifically oriented towards data statistical processing, for example to summarize the distribution of the observed values (central tendency, dispersion, range, etc.). The arithmetic mean of the values of x is obtained with the command mean():

```
> mean(x)
[1] 2604.9
```

Other commands, to which we will return in Chapter 2, make it possible for basic information to be obtained about the shape of the distribution of a continuous variable, such as the range or the variance, for example:

```
> range(x)
[1] 2523 2665
> c(min(x), max(x))
[1] 2523 2665
```

```
> var(x)
[1] 2376.767
```

As can be observed in the previous instructions, it is perfectly possible to combine two commands that return the same type of result, for example c(min(x), max(x)). As in the case of creating of a list of numbers, the command c() allows the results of commands returning numbers to be associated with the same list.

The following illustrations are based on the same idea as the division operation that is mentioned above: each operation (squaring of the values of x and centering these values on their average) is applied on a per-element basis.

```
> sum(x)
[1] 26049
> sum(x^2)
[1] 67876431
> sum(x^2 - mean(x))
[1] 67850382
```

In the previous example, mean(x) is a constant (which depends upon the data but that does not vary in this case) that we subtract from each x_i^2 (where i is the observation number or index in variable x), which amounts to 11 different values in total. In the expression x - x^2, on the other hand, the square of x is subtracted from each element of x (ten operations in total, with ten pairs of different numbers each time).

The commands sort(), order() and rank() allow the elements of a variable to be sorted or work with the ranks of the observations, that is to say with their position in the variable.

1.2.3. *Management of categorical variables*

Qualitative or categorical variables have their own distinct status in most statistical packages: their modalities are represented by numbers (1, 2, 3, etc.) but generally they are associated with identifiers, called labels in R. In the previous example, data about the weight of babies at birth (x) and the weight of their mother (y) were available. Suppose that it is also known if the mother was smoking during the first trimester of her pregnancy, and let this variable be called z. When the mother did not smoke during this period, the variable is equal to 1; when the mother was smoking, the variable is equal to 2. The augmented data of variable z are presented in Table 1.2.

x	2523	2551	2557	2594	2600	2622	2637	2637	2663	2665
y	82.7	70.5	47.7	49.1	48.6	56.4	53.6	46.8	55.9	51.4
z	1	1	2	2	2	1	1	1	2	2

Table 1.2. *Augmented artificial data about weights at birth*

The data will be entered in R as was done for the variables x and y:

```
> z <- c(1, 1, 2, 2, 2, 1, 1, 1, 2, 2)
> z
[1] 1 1 2 2 2 1 1 1 2 2
```

The factor(z) command allows this numerical variable be converted into a categorical variable. To associate labels to numerical codes, the labels= option is used. The levels= option enables the numerical codes of the levels of the variable to be specified. The value 1 will be associated with the NS modality and the value 2 to the S mode:

```
> factor(z, levels = c(1,2), labels=c("NS","S"))
 [1] NS NS S  S  S  NS NS NS S  S
Levels: NS S
```

When R displays the contents of variable z, it is effectively a variable of the factor type, with two levels, that are unordered here as NS and S. Some operations, such as the calculation of the arithmetic mean, do not generally make any sense with this type of variable and this is what also happens with R. Simple or crossed tabulation operations however remain entirely valid. Another aspect important: the labels associated with a qualitative variable can be recovered by using the levels command, while nlevels returns the total number of modalities:

```
> z <- factor(z, labels=c("NS","S"))
> levels(z)
[1] "NS" "S"
> nlevels(z)
[1] 2
```

1.2.4. *Manipulation of categorical variables*

It is often necessary to recode a k-class qualitative variable into $j < k$ classes or associate *labels* with the numerical modalities of a qualitative variable. For example, suppose that data on forced expiratory volume in 1 second (FEV1) (variable fev1) from ten patients are provided in the form of a three-tier ordered variable: critical

(1), low (2) and normal (3). Here, the `sample()` command will be used, which allows random data to be generated by resampling from a list of values. This is an R command which expects three options: the first shows the list of permissible values, the second the number of observations to be generated and the last the type of random draw that is sought: with (`replace = TRUE`) or without (`replace = FALSE`):

```
> fev1 <- sample(1:3, 10, replace=TRUE)
> fev1
 [1] 3 2 1 3 3 3 2 2 1 2
```

Note that the values are random and are not necessarily identical if using this command in an R session. To this end, one should use the `set.seed()` command to ensure the reproducibility of these simulations.

First, the numerical values will be replaced with the labels described previously (1 = critical, 2 = low and 3 = normal):

```
> fev1 <- factor(fev1, levels=c(1,2,3),
                  labels=c("critical","low","normal"))
> fev1
 [1] normal   low      critical normal   normal   normal   low
 [8] low critical low
Levels: critical low normal
```

The distribution of the counts by class can be verified quickly using the `table()` command which enables simple or cross-tabulations that will be discussed in Chapter 2:

```
> table(fev1)
fev1
critical      low   normal
       2        4        4
```

If the aim is to recode the three-class variable `fev1` into a two-class variable by aggregating the first two modalities or levels, the `levels()` command can be used as follows:

```
> levels(fev1)
[1] "critical" "low"        "normal"
> levels(fev1)[1:2] <- "critical or low"
> levels(fev1)
[1] "critical or low" "normal"
> table(fev1)
fev1
```

```
critical or low          normal
            6                4
```

It should be observed that the modalities of the variable fev1 have effectively been modified and the frequencies of the first two original levels can now be found in the first level as critical or low. It is advisable to always verify that the recoding operations of the levels of a categorical variable have been correctly carried out as expected, using the levels() or table() commands.

1.3. Selection of observations

1.3.1. *Index-based selection*

Considering the weight data discussed above, one could select more than one observation, for example the third and the fifth:

```
> x[c(3,5)]
[1]  2557 2600
```

or from the third to the fifth:

```
> x[3:5]
[1]  2557 2594 2600
```

The slightly peculiar syntax 3:5 actually designates the sequence of integers starting at 3 and ending at 5. This is, therefore, strictly equivalent to c(3,4,5). Thus to access specific values of a variable, it suffices to provide a list of observation numbers (or index). The same indexing principle applies of course to qualitative variables:

```
> fev1[2]
[1] critical or low
Levels: critical or low normal
```

1.3.2. *Criterion-based selection*

Suppose now that we want to obtain the weight of babies whose mother weighs less than 50 kg:

```
> x[y < 50]
[1]  2557 2594 2600 2637
```

The previous command reads: select the values of x for which the (logical) condition y< 50 holds. This condition can be directly visualized by typing y< 50 in R, which yields the following result:

```
> y < 50
[1] FALSE FALSE  TRUE   TRUE   TRUE FALSE FALSE  TRUE FALSE FALSE
```

This selection principle based on an external criterion remains valid when the criterion is a categorical variable. To obtain the list of weights of infants whose mother is a non-smoker, the following command should be entered:

```
> x[z == "NS"]
[1] 2523 2551 2622 2637 2637
```

The double equal sign (==) is being used to designate the logic condition "z equal to NS". Finally, it is possible to combine criteria relating to variables of the same type or of different types using the logical operators & (and) and | (or), mainly. The following command returns the values of x such as z is equal to NS *and* $y < 55$ (weight of babies whose mother does not smoke and who weight 55 kg or less):

```
> x[z == "NS" & y <= 55]
[1] 2637 2637
```

1.4. Representation and processing of missing values

In reality, it is rare for full data (without missing values) to be available. Regardless of the manner decided to statistically process sets, it is important to ensure that they are well represented as such by the statistical software.

Returning to the previous example, suppose that the third observation for the weight of babies is actually missing or that we want to process it as such. This datun is represented by a dot (.) in Table 1.3.

x	2523	2551	.	2594	2600	2622	2637	2637	2663	2665
y	82.7	70.5	47.7	49.1	48.6	56.4	53.6	46.8	55.9	51.4

Table 1.3. *Artificial data about weight at birth including the missing datum*

This missing value could have been represented by the lack of value (such as a blank cell in Microsoft Excel) or by any other symbol. Using the same approach as when variable x is set for the first time, a way of capturing the weight of the babies would be to replace the missing value with the term NA, which is the term reserved by R to encode the missing values:

```
> c(2523, 2551, NA, 2594, 2600, 2622, 2637, 2637, 2663, 2665)
```

Since variable x has already been entered, it can simply be updated, in this case replacing the third element with NA:

```
> x[3] <- NA
> x
[1] 2523 2551    NA 2594 2600 2622 2637 2637 2663 2665
```

Care should be taken, as the command length(x) always returns 10: there are effectively ten observations, but one of them is missing. The command is.na() makes it possible to verify the missing values in a variable:

```
> is.na(x)
[1] FALSE FALSE  TRUE FALSE FALSE FALSE FALSE FALSE FALSE FALSE
> which(is.na(x))
[1] 3
```

As seen in the case of the observation selection, the values returned by the command is.na() are Boolean values equal to true (TRUE) when the value of x is missing (NA), and false (FALSE) otherwise. This command will therefore return as many values as variable x contains. To facilitate the identification of the number of missing observations, it may be preferable to combine is.na() with the command which(). It should be noted that commands can be combined in an intuitive way: the which() command uses the result returned by is.na(), without having to create a dummy variable to store the result. An alternative consists of counting the so-called complete cases with the command complete.cases:

```
> complete.cases(x)
[1]  TRUE  TRUE FALSE  TRUE  TRUE  TRUE  TRUE  TRUE  TRUE  TRUE
```

Using the principle of observation selection used previously, the following command will thus return the values of y for which x has a non-missing value:

```
> y[complete.cases(x)]
[1] 82.7 70.5 49.1 48.6 56.4 53.6 46.8 55.9 51.4
```

1.5. Importing and storing data

1.5.1. *Univariate data*

The three sets of ten measures discussed in the preceding sections were of a sufficiently moderate size to allow their direct entry in R. However, in most cases, the data will have already been stored in an external file and the first step usually consists of importing them into R.

For example, here is the contents of the file `poids.dat`, which is a simple text file that can be visualized with any text editor:

```
2523 2551 2557 2594 2600 2622 2637 2637 2663 2665
```

In a simple case where only one set of measurements has to be imported, the `scan()` command is sufficient. It indicates the location of the data file. The term "location" designates the precise location of the file in the tree structure of the computer's file system. Here, it will be assumed that the working directory (editable with the `setwd()` command or using RStudio utilities) is the current directory:

```
> x <- scan("poids.dat")
Read 10 items
> head(x, n=3)
[1] 2523 2551 2557
```

The `head()` command enables the first observations of a variable to be displayed. The option n= enables their number to be specified (by default, n=6).

1.5.2. *Multivariate data*

In more complex, more realistic cases, data are multivariate with the variables generally organized in columns and the observations in lines [WIC 14]: each line of the file, therefore, represents a statistical unit in which several measurements or data points (variables or fields) have been collected, the latter being separated one from another by a comma, semicolons, spaces or tabs. In all cases, the command to use is `read.table()` (the field delimiter is a space or a tab) or one of its shortcuts: `read.csv()` (comma-type separator) and `read.csv2()` (semicolon-type separator).

Consider the file `birthwt.dat`, whose first three lines are:

```
0 19 182 2 0 0 0 1 0 2523
0 33 155 3 0 0 0 0 3 2551
0 20 105 1 1 0 0 0 1 2557
```

There are ten columns (that is ten variables) and each value is separated from the next by a space. The name of the variables does not appear anywhere, but we know that they are the following variables: weight status of the baby at birth `low` (= 1 if weight < 2.5 kg, 0 otherwise), `age` of the mother (years), `lwt` weight of the mother (in pounds), `race` ethnicity of the mother (encoded in three classes, 1 = white, 2 = black and 3 = other), `smoke` (= 1 if consumption of tobacco during pregnancy, 0 otherwise), ptl (number of previous premature labours), ht (= 1 if history of hypertension, 0 otherwise), ui (= 1 if manifestation of interuterine pain, 0 otherwise), ftv (number of consultations with a gynecologist during the first trimester of the pregnancy), bwt for the weight of the babies at birth (in grams). Here is how these data can be imported in R, always assuming that the data file is located in the working directory of R:

```
> bt <- read.table("birthwt.dat", header=FALSE)
> varnames <- c("low","age","lwt","race","smoke","ptl","ht",
                "ui","ftv","bwt")
> names(bt) <- varnames
> head(bt)
  low age lwt race smoke ptl ht ui ftv  bwt
1   0  19 182    2     0   0  0  1   0 2523
2   0  33 155    3     0   0  0  0   3 2551
3   0  20 105    1     1   0  0  0   1 2557
4   0  21 108    1     1   0  0  1   2 2594
5   0  18 107    1     1   0  0  1   0 2600
6   0  21 124    3     0   0  0  0   0 2622
```

If the data file had had the following form:

```
0,19,182,2,0,0,0,1,0,2523
0,33,155,3,0,0,0,0,3,2551
0,20,105,1,1,0,0,0,1,2557
```

which is typically the export format offered by a spreadsheet program such as Excel, then the command read.table() could have simply been replaced by read.csv() (or change the option sep= of read.table()).

1.5.3. *Storing the data in an external file*

In the univariate case as in the multivariate case, the data processed in R can be exported using write.table() or write.csv() to create text files identical to those discussed above. For example:

```
> write.csv(bt, file="bt.csv")
```

will save the file imported with the read.table() command in the form of a file where the values of the variables are separated by commas and it will be possible to open the bt.csv file with a spreadsheet program such as Microsoft Excel.

If data must be directly saved in the R format, save(bt, file="bt.RData"), will have to be used, RData (or rda) constituting the extension reserved for R database files.

1.6. Multidimensional data management

1.6.1. *Construction of a structured data table*

So far, we have created and manipulated different variables (x , y and z, mainly). A more natural way to account for this association between observations is to gather these variables inside the same data structure or table, called a "data frame" in R.

The general structure of a data frame is illustrated in Figure 1.1. The hypothetical data are expected to be gathered in a data frame named a. Data are collected in seven statistical units, arranged in lines, and the four variables correspond to a numerical score (score), the gender of the individual (gender), intelligence quotient (IQ) and socio-economic level (SES). A data frame is thus a rectangular table enabling the presentation of variables of different types (numbers, characters, factors, etc.) but of the same size in columns. A row will designate a statistical unit. The first unit thus has a score of 1, it is of male gender, has an IQ of 92 and its socio-economic status is C. The columns of a data frame may be named, which allows to be accessed the variables by their name.

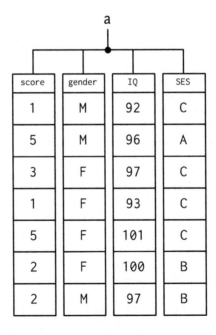

a

score	gender	IQ	SES
1	M	92	C
5	M	96	A
3	F	97	C
1	F	93	C
5	F	101	C
2	F	100	B
2	M	97	B

Figure 1.1. *Structure of a data frame*

This principle can be applied to the variables x, y and z used in the preceding sections. These three variables can be stored in three separate columns by using the command data.frame(). Here is how one can proceed in R:

```
> d <- data.frame(x, y, z)
> names(d) <- c("weight.baby", "weight.mother", "cig.mother")
> head(d, n=3)
  weight.baby weight.mother cig.mother
1        2523          82.7         NS
2        2551          70.5         NS
3        2557          47.7          S
```

It should be noted that we took the opportunity to change the name of the original variables. The first individual therefore weighs 2,523 g and her mother who weighs 82.7 kg does not smoke. To display only this information, the table, named d, will be indexed by the corresponding row number, for example:

```
> d[1,]
  weight.baby weight.mother cig.mother
1        2523          82.7         NS
```

If nothing is specified for the column numbers, it means that one wishes to select them all. In contrast, to display the value of the second variable for the first two observations, we will use:

```
> d[c(1,2), 2]
[1] 82.7 70.5
```

Instead of using the number of columns, it is perfectly possible to use the name of the variable. For example, a statement such as d$weight.baby will return all values of the weight.baby variable, so that d$weight.mother[1:2] would return in the first two observations of the variable weight.mother as in the previous case.

1.6.2. *Birthwt data*

Comprehensive data on the weight of the babies discussed previously are included among the basic examples provided with R and they can be imported using the data() command. The data come from an epidemiological study carried out in the 1980s [HOS 89] to assess the risk factors relating to low birth weight American children. This data set will be used systematically in the following chapters, with the exception of that relating to the analysis of survival data (Chapter 7). Once imported, this data table is available under the name birthwt:

```
> data(birthwt, package="MASS")
> c(nrow(birthwt), ncol(birthwt))
[1] 189  10
> names(birthwt)
 [1] "low"   "age"   "lwt"   "race"  "smoke" "ptl"   "ht"
 [8] "ui"    "ftv"   "bwt"
> head(birthwt, n=2)
   low age lwt race smoke ptl ht ui ftv  bwt
85   0  19 182    2     0   0  0  1   0 2523
86   0  33 155    3     0   0  0  0   3 2551
```

The birthwt variable is a table (data frame) comprising ten variables, each comprising 189 observations. Therefore, information on 189 statistical units are available in this prospective study. To access the values of a particular variable, the following notation will be used: table name (birthwt) followed by the name of the variable prefixed with the dollar ($) sign. For example, the first five values for the weight of babies (bwt) are thus:

```
> birthwt$bwt[1:5]
[1] 2523 2551 2557 2594 2600
```

The meaning of each variable is given above. It is known that some variables are strictly binary (0/1), such as low, smoke, ht and ui, while the other variables are numeric, either with discrete values such as ftv or with values assumed to be continuous such as lwt or bwt. We know that in the case of binary variables, a value of 1 means that the sign is present (the mother smokes, the mother has a history of high blood pressure, etc.). It costs nothing to add more informative labels to these variables. Ethnicity (race) is more specific because this is a qualitative variable but is actually processed by R as a numerical variable:

```
> summary(birthwt$race)
   Min. 1st Qu.  Median    Mean 3rd Qu.    Max.
  1.000   1.000   1.000   1.847   3.000   3.000
```

The summary() command allows the distribution of numerical and categorical variables to be summarized, and applies equally to a single variable as to a data table. A way to recode the modalities of the qualitative variables for the birthwt table is as follows:

```
> yesno <- c("No", "Yes")
> birthwt$smoke <- factor(birthwt$smoke, labels=yesno)
> birthwt$race <- factor(birthwt$race, levels=c(1, 2, 3),
                    labels=c("White", "Black", "Other"))
```

The same approach will be used for low, ht, ui (use the labels yesno), and it shall be verified that the final data are now in the expected format using summary(), which makes it possible to quickly identify the variables considered numerical (numerical summary with min, max, mean) or qualitative (values table):

```
> birthwt <- within(birthwt, {
    low <- factor(low, labels=yesno)
    ht <- factor(ht, labels=yesno)
    ui <- factor(ui, labels=yesno)
})
```

Rather than repeating the name of the data frame followed by the name of the variable each time, operations have been regrouped using the within() command. It is unnecessary to specify the numeric codes (levels=) for these three variables since they are all binary.

At this point, here is what the recoding of variables gives when using the summary() command with the first five variables:

```
> summary(birthwt[,1:5])
     low           age              lwt              race        smoke
 No :130    Min.    :14.00    Min.    : 80.0    White:96    No :115
 Yes: 59    1st Qu.:19.00    1st Qu.:110.0    Black:26    Yes: 74
            Median :23.00    Median :121.0    Other:67
            Mean    :23.24    Mean    :129.8
            3rd Qu.:26.00    3rd Qu.:140.0
            Max.    :45.00    Max.    :250.0
```

It is now possible to answer the following questions: what is the average weight of women who smoked during their pregnancy? How many cases of hypertension can be identified in women weighing more than 60 kg (knowing that the measurements in the table are expressed in pounds)? What is the minimum weight of babies with mothers who have not demonstrated interuterine pain? The following is a possible solution for each of these questions:

```
> mean(birthwt$lwt[birthwt$smoke == "Yes"]) ## weight in pounds
[1] 128.1351
> table(birthwt$ht[birthwt$lwt/2.2 > 60])    ## weight in kg

No Yes
55   7
> min(birthwt$bwt[birthwt$ui == "No"])
[1] 1135
```

Subsequently, rather than repeating the name of the data frame followed by $ to access one or more variables, as in the previous example, it is preferable to use the with() command that makes it possible to indicate in which data frame the search for variables is to be carried out:

```
> with(birthwt, lwt[1:5])
[1] 182 155 105 108 107
> with(birthwt, mean(lwt[smoke == "Yes"]))
[1] 128.1351
```

1.7. Key points

– A variable will be represented in the form of a list of numbers or characters (single or multiple).

– The variables will generally be arranged in columns in a rectangular array (data frame).

– The usual arithmetic operations operate on each element (observation) of a variable.

– It is possible to index observations by their number or position in a list of values, or by using logical expressions.

1.8. Going further

The book by Phil Spector [SPE 08] provides additional information about importing external data sources (including the case of relational databases) merging of multiple data sources with the merge() command, handling strings and the representation of dates in R. Other works are available online, free of cost or for a moderate price, for example *R Programming for Data Science* by Roger Peng (https://leanpub.com/rprogramming) or *The Elements of Data Analytic Style* by Jeff Leek (https://leanpub.com/datastyle).

1.9. Applications

1) Plasma viral load is used to describe the viral load (for example, HIV) in a blood sample. This viral marker that allows the progression of infection and effectiveness of treatments to be measured where the number of copies per milliliter are recorded and most measurement instruments have a detectability threshold of 50 copies/ml. Here follows a series of measurements, X, expressed in logarithms (base 10), collected on 20 patients:

```
3.64 2.27 1.43 1.77 4.62 3.04 1.01 2.14 3.02 5.62 5.51 5.51 1.01
1.05 4.19 2.63 4.34 4.85 4.02 5.92
```

As a reminder, a viral load of 100,000 copies/ml is equivalent to 5 log.

We want to answer the following:

a) Indicate how many patients have a viral load considered as non-detectable.

b) The researcher realizes that the value 3.04 corresponds to a data entry error and must be changed to 3.64. Similarly, she has a doubt about the seventh measurement and decides to consider it as a missing value. Perform the corresponding transformations.

c) What is the median viral load level in copies/ml, for the data considered valid?

First and foremost, it is necessary to express the detection limit (50 copies/ml) in logarithmic units; this is, in fact, equal to:

```
> log10(50)
```

Next, we need to filter the observations that do not verify the condition $X > 1.70$ (the exact numeric result will be used, not the approximate value):

```
> X <- c(3.64,2.27,1.43,1.77,4.62,3.04,1.01,2.14,3.02,5.62,5.51,
        5.51,1.01,1.05,4.19, 2.63,4.34,4.85,4.02,5.92)
> length(X[X <= log10(50)])
```

To replace the observation equal to 3.04 we will perform a simple logic test so this observation is not duplicated in the observations series:

```
> X[X == 3.04] <- 3.64
```

Concerning the seventh observation, we will proceed in the same manner by updating the value of the observation with the corresponding index:

```
> X[7] <- NA
```

Finally, the median viral load for patients with a measurement considered valid can be calculated as follows:

```
> Xc <- X[X > log10(50)]
> round(median(10^Xc), 0)
```

2) The dosage.txt file contains a series of 15 bioassays, stored in numerical format with three decimal places as follows:

```
6.379 6.683 5.120 ...
```

– use scan to read these data (thoroughly read the online help regarding the use of this command, particularly the what= option);

– correct the series of measurements to calculate the arithmetic average;

– store the corrected data in a text file called data.txt.

When the data stored in a text file are not too numerous, it is recommended they are double checked using a simple text editor before importing into R. In this case, it turns out that there has been an encoding problem with the decimal separator: in one of the measurements, a comma was used instead of a dot to separate the integer part from the fractional part of the numbers. Considering the following statement:

```
> x <- scan("dosage.txt")
```

R will return the error message:

```
Error in scan(file, what, nmax, sep, dec, quote, skip, nlines):
  scan() was expecting "a real" and "2,914" was entered
```

since by default R tries to read numbers (in English format). To correct this, it is necessary that the format in which you want to store the data be specified.

```
> x <- scan("dosage.txt", what="character")
> str(x)
> head(x)
```

However, it is not possible to directly calculate the average of the data imported in this way because they are in the form of characters (character). Therefore, the erroneously recorded observation will be corrected, before the measurements are converted into numbers employable by R:

```
> x[x == "2,914"] <- "2.914"
> x <- as.numeric(x)
> head(x)
> round(mean(x), 3)
```

Finally, to save the data in a new file, one can use the general write.table() command, which includes several options, but that generally works as follows:

```
> write.table(x, file="data.txt", row.names=FALSE)
```

An overview of the text file saved in this manner is provided below:

```
"x"
6.379
6.683
5.12
```

```
6.707
6.149
5.06
```

If the name of the variable does not have to appear on the first line, the option `col.names=FALSE` can be added.

3) The `anorexia.dat` file contains data from a clinical study in anorexic patients who received one of the three following therapies: behavioral therapy, family therapy, control therapy [HAN 93].

a) How many patients are there in total? How many patients are there per treatment group?

b) The weight measures are in pounds. Convert them into kilograms.

c) Create a new variable containing difference scores (`After - Before`).

d) Indicate the mean and the range (min/max) of the difference scores per treatment group?

An overview of the data contained in the `anorexia.dat` file is provided below (five first lines of the file):

```
Group Before After
g1 80.5  82.2
g1 84.9  85.6
g1 81.5  81.4
g1 82.6  81.9
```

The first line is a header row indicating the name of the variables, and each of the following lines represents a statistical unit for which the identity group, the weight measurement before and the weight measurement after the therapy are available. To import this data file, we will use the `read.table()` command by specifying the `header=TRUE` option to take the header row into account properly:

```
> anorex <- read.table("anorexia.dat", header=TRUE)
> names(anorex)
> head(anorex)
```

The total number of patients corresponds to the number of rows in the data table:

```
> nrow(anorex)
```

There are, therefore, 72 patients. The easiest way to find the distribution of patients by group, is to perform a simple tabulation of the qualitative variable `Group`:

```
> table(anorex$Group)
```

To convert weight expressed in pounds into kilograms, we can divide each measurement by 2.2 (approximately). This operation is performed for each of the two variables Before and After:

```
> anorex$Before <- anorex$Before/2.2
> anorex$After <- anorex$After/2.2
```

It should be noted that in this case the original values of these two variables will simply be replaced. To access then again, it will be necessary to reload the file. Another solution would have been to create two new variables:

```
> anorex$Before.kg <- anorex$Before/2.2
> anorex$After.kg <- anorex$After/2.2
```

To calculate the After-Before difference scores, we can subtract the values of the two variables, remembering that this kind of operations is carried out on a per element basis (for each statistical unit). Here, the newly created variable will be added in the anorex data table:

```
> anorex$weight.diff <- anorex$After - anorex$Before
> head(anorex)
```

Finally, to calculate the average and the range of the difference scores by treatment group, we can proceed in two ways: either each group is isolated and the requested statistics are computed; or the computing operation is done for each modality of the qualitative variable. For example the first solution, for the first group would be obtained, using:

```
> mean(anorex$poids.diff[anorex$Group == "g1"])
> range(anorex$poids.diff[anorex$Group == "g1"])
```

2

Descriptive Statistics and Estimation

From this chapter and up to Chapter 6, on logistic regression, we will work with the same data set containing the weights at birth [HOS 89]. The objective of this chapter is to present the tools available in R for the univariate description of a numerical or a categorical variable. The summary of a variable can be done numerically by using central tendency and dispersion indicators for a numeric variable or a table of counts or frequencies for a categorical variable. The distribution of the observed counts for a variable can also be visualized in the form of a graph, either by using the individual data points (univariate scatterplot) or by using aggregated data (histogram, box plot, bar chart and line plot).

2.1. Summarizing a numerical variable

2.1.1. *Central tendency and shape of the distribution*

Consider the variable bwt, which represents the weight of newborns. The weights are expressed in grams and there are no missing counts. This can quickly be verified by counting the number of frequencies for which the is.na() command returns true (TRUE):

```
> sum(is.na(birthwt$bwt))
[1] 0
```

In the previous expression, two commands are combined: the is.na() command returns the value TRUE if one of the elements of bwt is considered to be a missing value, it returns the value otherwise FALSE. The result of this command, therefore, contains as many elements as the variable bwt and the number of counts equal to TRUE is determined, these being represented internally as numbers with the value 1 (hence, the use of the command sum()).

The main indicators of central tendency, mean and median, can be obtained with the mean() and median() commands. They assume that there are no missing values, which is the case here; if so it is necessary to add the option na.rm=TRUE.

```
> mean(birthwt$bwt)
[1] 2944.587
> median(birthwt$bwt)
[1] 2977
```

As seen in the previous chapter, the summary() command can also be used directly. This has the advantage of displaying the range of the observed frequencies (minimum and maximum, see min(), max() and range() commands) and the quartiles (the median corresponding to the second quartile):

```
> summary(birthwt$bwt)
   Min. 1st Qu.  Median    Mean 3rd Qu.    Max.
    709    2414    2977    2945    3487    4990
```

The quartiles can be restored as follows:

```
> quantile(birthwt$bwt)
   0%   25%   50%   75%  100%
  709  2414  2977  3487  4990
```

The following statement:

```
> quantile(birthwt$bwt, probs=seq(0.1, 1, by=0.1))
```

returns the value of deciles of the bwt variable. It should be noted that instead of writing each decile (10%, 20%, etc.), it is preferable to let R generate the sequence of deciles using the seq() command.

With regard to the estimation of asymmetry or kurtosis of the distribution, there are no base commands in R, but they are available in the e1071 (skewness() and kurtosis()) package. This package must be installed and loaded before these commands are used.

2.1.2. Distribution indicators

Concerning distribution measures, quartiles are already available facilitating the estimation of the interquartile range (50% of observations). Therefore, this interval can effectively be calculated as follows:

```
> quantile(birthwt$bwt, probs=c(0.75))
  - quantile(birthwt$bwt, probs=c(0.25))
75%
1073
```

or more simply

```
> diff(quantile(birthwt$bwt, probs=c(0.25, 0.75)))
75%
1073
```

since the diff() command returns the iterated differences of the counts of a variable. The reason why 75\% is displayed above the value of the interquartile range is due to the fact that R explicitly "labels" the quantiles as observed using quantile(birthwt$bwt). An alternative would consist of directly using the IQR() command.

R also provides an estimator of the standard deviation and the variance (with an $n - 1$ denominator) by means of the sd() and var() commands.

```
> var(birthwt$bwt)
[1] 531753.5
> sd(birthwt$bwt)
[1] 729.2143
> sd(birthwt$bwt)^2
[1] 531753.5
```

The same precautions of usage concerning missing frequencies apply and it will be necessary to add the na.rm=TRUE option in the presence of missing data.

2.2. Summarizing a categorical variable

Regarding the qualitative variables, whether the terms of the variable are ordered or not, a table of counts or frequencies can always be provided. The table() command which was used in the previous chapter makes it possible to establish a frequency table. Similarly to the mean() or median() commands, it is important to specify what has to be done in the presence of missing counts. By default, R will not display them as a separate category; if the missing observations have to be displayed explicitly, the useNA="always" option has to be added:

```
> table(birthwt$race)

White Black Other
   96    26    67
```

It is perfectly possible to save the intermediate results, including the output from commands such as `table()`, in R variables. This facilitates their reuse and improves the readability of the commands using these same results. In the following example, the use of `table()` will be combined with `prop.table()` to establish a relative frequencies table rather than a table of counts:

```
> tab <- table(birthwt$race)
> tab

White Black Other
  96    26    67
> prop.table(tab)

    White       Black       Other
0.5079365 0.1375661 0.3544974
```

We could add * 100 to the preceding expression to express the results in percentages rather than in relative frequencies. Note that the multiplication operation is carried out element-wise, as was seen in the case of operations on numeric variables (section 1.2.2).

The mode can be found from a table of figures by searching for the highest value, as illustrated hereafter:

```
> which.max(tab)
White
    1
```

Here, tab represents the table constructed using `table(birthwt$race)` correctly. The result returned by `which.max()` tells us that the modal class is the first class of the `race` variable (White, which appears in the first position in the list of the modalities of the variable, see `levels(birthwt$race)`).

There is an another R command that allows the same type of operation (simple tabulation and cross tabulation, the latter being presented in Chapter 3), `xtabs()`:

```
> xtabs(~ birthwt$race)
birthwt$race
White Black Other
  96    26    67
```

This command involves a new notation, the use of the symbol ~, which is used to describe relationships between variables, for example to express a relation of numeric variable, x, described by categorical variable z (x ~ z). In the univariate case, the notation ~ birthwt$race instructs R to describe the variations of the race variable of the birthwt table, which for a qualitative variable is equivalent to the distribution of the frequencies by class or by variable mode. This notation is called "formula notation" and will be covered again in the following chapters.

In some situations, it might be desirable to summarize a numeric variable using a frequency table considering different class intervals (upper and lower bounds between which the counts of the numeric variable range): we are then left with a categorical variable with which table() or xtabs() can be used. Consider the age of the mothers (age, which varies between 14 and 45 years) and the following class intervals: $[15, 19]$, $[20, 24]$, $[25, 29]$, $[30, 45]$. As they are defined, these intervals include the counts indicated for their upper and lower bounds. A way to build the frequency table corresponding to this distribution of individuals in mutually exclusive age classes is presented as follows:

```
> table(cut(birthwt$age, breaks=c(15,19,24,29,45),
            include.lowest=TRUE))

[15,19] (19,24] (24,29] (29,45]
     48      69      42      27
```

It should be noted that the use of the cut() command requires that R be instructed about how the counts of the numeric variable are to be split and, in particular, if the lower or upper bounds need to be inclusive. This example indicates that the intervals must include the lower bound of the first interval (include.lowest=TRUE). Another useful option is right=; it is used to indicate whether the intervals should be open on the right (by default, R defines intervals which bounds are opened on the left and closed on the right).

2.3. Graphically representing the distribution of a variable

All graphics commands used in the book are provided by the lattice package, which is available with the basic installation of the R software. There are also basic graphics commands, but as lattice provides graphics whose default options are relatively satisfactory, it will be used preferrably. An introduction to the lattice package is provided in Appendix 2.

To use these commands, it is necessary to type:

```
> library(lattice)
```

before using any of the commands in the package. It is not necessary to repeat the operation as long as the R session in progress is not closed down.

2.3.1. *The case of numerical variables*

To graphically represent all of the observations, a simple one-dimensional scattergraph can be employed using `stripplot()`. An example of its usage is provided hereafter with a variable that is assumed to be called `weight`. When the number of observations is not too high, this allows us to quickly visualize extreme data as well as the central tendency:

```
> stripplot(~ weight)
```

Once again, the formula notation presented above comes across: here, in a univariate context, R is given instructions to describe the variations in the `weight` variable, which for a numeric variable is equivalent to the distribution of all of the observations, without consideration of any classification or stratification factor. It is also possible to use this type of graphic representation with large data sets, but there is a risk in this case that points may overlap, making the interpretation of the univariate distribution harder. An alternative consists of adding a slight offset, horizontal or vertical depending on the axis selected to represent the variable measured, to the points on the graph. This is illustrated in Figure 2.1 with the weight of babies:

```
> stripplot(~ bwt, data=birthwt, jitter.data=TRUE, amount=.05,
        xlab="Weight (g)")
```

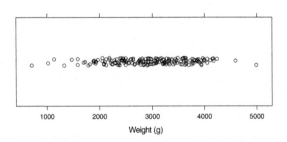

Figure 2.1. *Dotchart for the distribution of the weight of the babies*

The vertical offset of the points is controlled by the `jitter.data=TRUE` option (in fact, a Gaussian random variate is added for each point on the y-axis) and the amplitude of the offset can be increased with the `amount=` option.

A well-known representation for continuous variables is a counts or frequencies histogram by defining a series of class intervals for the counts taken for the numeric variable. R offers different options for the choice of the calculation of the class intervals and they are generally satisfactory. It is always possible to manually specify the number of intervals (`nint=` option) or the classes themselves (`breaks=` option). For example, to distribute the frequencies over intervals of 500, an option such as `breaks=seq(700, 5000, by=500)` could be employed, the bounds 700 and 5,000 being slightly outside the range of frequencies observed for the `bwt` variable. By default, R displays a proportions histogram that can be changed by adding the `type="count"` option to obtain a histogram of the counts. The following gives the result illustrated in Figure 2.2 for the variable `bwt`:

```
> histogram(~ bwt, data=birthwt, type="count", xlab="Weight (g)",
            ylab="Values")
```

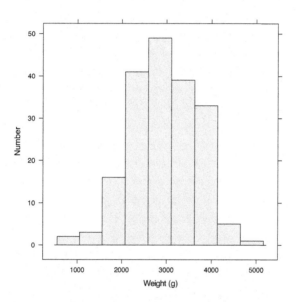

Figure 2.2. *Histogram for the distribution of the weight of the babies*

Here, an important concept is introduced in R: numerous commands provide the possibility of designating the data frame in which the variables are located by using the `data=` option, and directly using the name of variables rather than prefixing them systematically with the name of the table followed by the character $. Along with the formula notation, these two concepts form the basis of the description of statistical data in R. This approach can also be used to consider a subset of the observations of the data table specified by the `data=` option. To this end, we can add the `subset=` option and indicate a logical condition allowing the filtering of statistical units. This filter can be customized for the variable of interest; for example `subset = bwt >`

3000 (consider infants weighing more than 3 kg only) or any variable (or combination of variables by means of the logical operators & and |) available in the data frame, for example smoke == "Yes" (babies whose mother smokes):

```
> histogram(~ bwt, data=birthwt, subset=smoke == "Yes",
            type="count")
```

An alternative to the histogram is the non-parametric density curve that allows us to overcome the problem of the choice of class intervals. In this case, rather than regrouping the observations in pre-defined classes (manually or using a specific algorithm), it is perfectly possible to use more sophisticated methods to estimate the density curve, controlling the "smoothing" degree of this approximation, as illustrated in Figure 2.3:

```
> densityplot(~ bwt, data=birthwt, xlab="Weight (g)",
              ylab="Density")
```

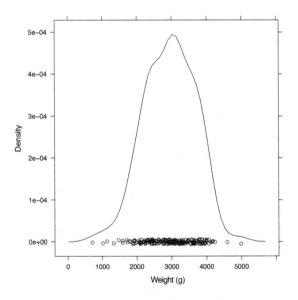

Figure 2.3. *Density curve estimated for the distribution of the weight of the babies*

The size of the smoothing window of the curve is controlled with the bw= option; the larger it is, the more attenuated the local variations are (for example, with bw=1000 the curve appears much closer to a Gaussian curve for the weight of the babies). The online help for the density() command provides additional information on the various parameters that can be controlled. Regarding lattice specifically, the corresponding online help is obtained with:

help(panel.densityplot)

Finally, we can use box-plot type diagram (Figure 2.4). In this type of representation, the "box" contains 50% of the observations so that these two edges represent the first and the third quartiles. The median is represented by a point inside this box. The "whiskers" represent the minimum and the maximum frequency of the distribution, except in the case where certain observations are located at more than 1.5 × IQR (IQR = interquartile range) from the first or from the third quartile, in which case these latter are indicated explicitly on the chart. The "whiskers" then represent the bounds limit frequencies (1.5 × IQR of the first or the third quartile, Tukey's method, see help(boxplot.stats)):

> bwplot(~ bwt, data=birthwt, xlab="Weight (g)")

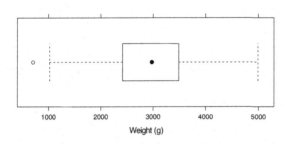

Figure 2.4. *Boxplot type representation of the distribution of the weight of the babies*

2.3.2. *The case of categorical variables*

In the case of categorical variables, we will either use bar graphs or line plots (Cleveland diagrams), avoiding circular representations of the "pie chart" type as much as possible.

In both cases, it is necessary to work with tabulated data, that is a table of figures constructed using table() or xtabs(). These commands can also be associated with prop.table() to obtain a table of relative frequencies. As seen above, the xtabs() command allows a formula notation which makes it possible to highlight the role played by the variables and to directly designate the data frame in which the variables in question are located. Here follows an example of a bar diagram for a single variable (Figure 2.5):

```
> barchart(xtabs(~ race, data=birthwt), xlab="Ethnicity",
        ylab = "Values", horizontal = FALSE)
```

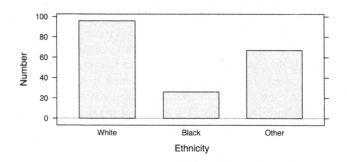

Figure 2.5. *Barchart for the distribution of frequencies according to the ethnicity of the mother*

It is possible to change the orientation of the bars (horizontal by default) using the `horizontal=FALSE` option (vertical bars). Dot plots, as shown in Figure 2.6, are built exactly the same way, with the `dotplot()` command that shares a large portion of the basic `barchart()` command option (in particular the `horizontal=FALSE` option for a vertically oriented diagram):

```
> dotplot(xtabs(~ race, data=birthwt), ylab="Ethnicity",
        xlab = "Values")
```

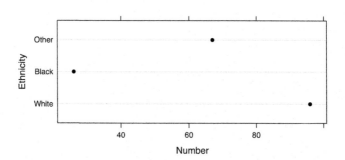

Figure 2.6. *Dot plot for the distribution of frequencies according to the ethnicity of the mother*

2.4. Interval estimation for a mean or a proportion

2.4.1. *Confidence interval for a mean*

The mean() and summary() commands both return an estimate of the average (arithmetic mean) for a numeric variable. To obtain $100(1 - \alpha)\%$ confidence intervals, where α refers to the Type I error risk (usually 5%) and referring to a Gaussian distribution (normal distribution), it is necessary to have an estimate of the standard error and to know the value of the quantile of the corresponding standard normal distribution. With regard to the latter, rather than using the approximation ± 1.96 (for $\alpha = 5\%$, or 2.5% for each distribution tail), its value can be obtained directly in R using the qnorm() command which provides the fractiles of a standard normal distribution $\mathcal{N}(0; 1)$:

```
> qnorm(0.975)
[1] 1.959964
```

It can be verified that the 1-pnorm(1.96) command returns the area under the density curve for a distribution tail on the right.

The standard error, defined as the ratio between the standard deviation estimated with the data and the square root of the sample size, will be obtained using:

```
> sd(birthwt$bwt) / sqrt(length(birthwt$bwt))
[1] 53.04254
```

As it has been seen in the previous chapter, rather than systematically repeating the name of the data frame followed by the symbol $ to work with a given variable, we can replace the previous expression with:

```
> with(birthwt, sd(bwt) / sqrt(length(bwt)))
```

which instructs R to work with the bwt variable available in the birthwt table. There are no missing frequencies in this data set and, therefore, the length() command returns the size of the sample correctly. Conversely, sqrt(sum(!is.na(birthwt$bwt))) should be used. By combining the two previous results, the lower and the upper bounds of a 95% confidence interval can be obtained for the average of the weights of the babies in this sample:

```
> mean(birthwt$bwt) - qnorm(0.975) * sd(birthwt$bwt) /
    sqrt(length(birthwt$bwt))
[1] 2840.626
> mean(birthwt$bwt) + qnorm(0.975) * sd(birthwt$bwt) /
    sqrt(length(birthwt$bwt))
[1] 3048.549
```

These two results can be stored in auxiliary variables and combined to facilitate reading:

```
> bwt.lci <- mean(birthwt$bwt) - qnorm(0.975) * sd(birthwt$bwt) /
    sqrt(length(birthwt$bwt))
> bwt.uci <- mean(birthwt$bwt) + qnorm(0.975) * sd(birthwt$bwt) /
    sqrt(length(birthwt$bwt))
> c(bwt.lci, bwt.uci)
[1] 2840.626 3048.549
```

Note that the calculation of the two bounds of the confidence interval could have been combined in a single command:

```
> mean(birthwt$bwt) + c(-1,1) * qnorm(0.975) * sd(birthwt$bwt) /
    sqrt(length(birthwt$bwt))
```

Furthermore, rather than calculating these confidence intervals manually, it is possible to use the epiR package, which contains a specific command for this type of calculation: epi.conf().

In the case of small samples, it may be preferable to use a Student's T-test to build the confidence interval associated with an average. The pt() and qt() command can be used instead of pnorm() and qnorm() to obtain the probability or the fractile counts t for ν degrees of freedom. The degrees of freedom for the student distribution are controlled by the df= option. The fractile associated with a probability of 0.025 on the right of a 18 degree of freedom distribution is thus:

```
> qt(0.975, df=18)
[1] 2.100922
```

An alternative way to obtain this type of confidence interval is to use a test procedure directly, t.test() for example, which returns the result of a hypothesis test and provides the confidence interval associated with the test statistic. In this case, a one-sample Student's t-test would be used, considering $H_0 : \mu = 0$ as the null hypothesis.

2.4.2. Confidence interval for a proportion

By using a normal approximation for the estimation of the confidence interval of a proportion, we would proceed in exactly the same manner as that described before, by replacing the estimates of the mean by that of the frequency p (obtained from a command like `table()`, for example) and of the standard error by $\sqrt{p(1-p)/n}$. As in the previous case this approximation is essentially valid for large samples.

With the data on the smoking status of mothers (`smoke` variable), a 95% confidence interval for the proportion of mothers smoking during pregnancy can, therefore, be formulated as:

```
> tab <- prop.table(table(birthwt$smoke))
> p <- tab[2]
> n <- nrow(birthwt)
> p + c(-1,1) * qnorm(0.975) * sqrt((p*(1-p))/n)
[1] 0.3219487 0.4611201
```

The previous sequence of commands makes it possible to illustrate some of the concepts mentioned in Chapter 1: it is possible to store the result of a `table()` or `prop.table()` command in an auxiliary variable. Here it is a simple tabulation of the smoke variable and, therefore, the p variable contains two counts:

```
> tab

      No        Yes
0.6084656 0.3915344
```

the second corresponding to the proportion of mothers who smoke: `tab[2]`, therefore, allows us to access this value. An alternative would be to use the syntax `tab["Yes"]` (the column name instead of the column number). There are no missing data, so `length(birthwt$bwt)` and `nrow(birthwt)` will essentially return the same information: the size of the sample. As a result, this quantity can be associated with a variable called n. The remainder corresponds to the expression of the confidence interval $p \pm 1.96\sqrt{p(1-p)/n}$ in the R language.

2.5. Key points

– The `summary()` command provides most of the information regarding the shape of the distribution of a numeric variable and a frequency table for the categorical variables. Most of this information can be obtained individually with specific commands such as `mean()`, `sd()`, `quantile()` or `table()`.

– A formula notation is used to describe variations of a variable and it serves as a point of entry for the graphics base commands to represent the distribution of a numerical or categorical variable.

– To construct asymptotic confidence intervals for an average or a proportion estimated from observed data, their mathematical formulation can be translated in the

R language starting from the qnorm() command and the calculation of the standard error directly associated with these quantities.

2.6. Applications

1) A quantitative variable X takes the following frequencies for a sample of 26 subjects:

24.9,25.0,25.0,25.1,25.2,25.2,25.3,25.3,25.3,25.4,25.4,25.4,25.4,
25.5,25.5,25.5,25.5,25.6,25.6,25.6,25.7,25.7,25.8,25.8,25.9,26.0

a) calculate the mean, the median and the mode of X;

b) what is the value of the variance estimated from these data?

c) assuming that data are grouped into four classes whose bounds are: 24.9-25.1, 25.2-25.4, 25.5-25.7, 25.8-26.0, display the distribution of the figures by class in the form of a table of counts;

d) represent the distribution of X as a histogram, without consideration of *a priori* class intervals.

One way to represent this type of data consists of entering the observations like this:

```
> x <- c(24.9,25.0,25.0,25.1,25.2,25.2,25.3,25.3,25.3,25.4,25.4,
         25.4,25.4,25.5,25.5,25.5,25.5,25.6,25.6,25.6,25.7,25.7,
         25.8,25.8,25.9,26.0)
```

The median and the average are obtained by using the mean() and median() commands, or directly via the summary() command:

```
> median(x)
> mean(x)
```

It can be verified that the median is the value correctly corresponding to the second quartile or to the 50th percentile that is to say the value of X such that 50% of the observations are smaller:

```
> quantile(x)
```

Regarding the mode, it is necessary to display the distribution of the figures according to the frequencies of X, then to verify what value of X is associated with the greatest count:

```
> table(x)
> names(table(x)[table(x)==max(table(x))])
```

In this particular case, it can be verified that two modes are found.

The variance is obtained using the var() command:

```
> var(x)
```

Assuming that the data are grouped into four classes whose bounds are: 24.9-25.1, 25.2-25.4, 25.5-25.7, 25.8-26.0, it is possible to recalculate the distribution of the sample by class:

```
> xc <- cut(x, breaks=c(24.9,25.2,25.5,25.8,26.0),
            include.lowest=TRUE, right=FALSE)
> table(xc)
```

It should be noted that R uses the English notation to represent the bounds of the class intervals: the symbol) on the right of a number means that this number is excluded from the interval, while] means that the interval contains this number. By default, R automatically determines the class intervals of a histogram produced when using the histogram() command from the lattice package. If we want to specify the bounds of the class intervals, the breaks= option will be used. For example, to display the distribution of the frequencies according to the four classes defined above, we would write:

```
> histogram(~ x,type="count",breaks=c(24.9,25.2,25.5,25.8,26.0))
```

2) The survival time of 43 patients suffering from chronic granulocytic leukemia, measured in days since the diagnosis [EVE 01] is available:

```
7,47,58,74,177,232,273,285,317,429,440,445,455,468,495,497,532,
571,579,581,650,702,715,779,881,900,930,968,1077,1109,1314,1334,
1367,1534,1712,1784,1877,1886,2045,2056,2260,2429,2509
```

a) calculate the median survival time;

b) how many patients have a survival (strictly) lower than 900 days at the time of the study?

c) what is the duration of survival associated with the 90th percentile?

The data are entered manually, as in the previous case:

```
> s <- c(7,47,58,74,177,232,273,285,317,429,440,445,455,468,495,
         497,532,571,579,581,650,702,715,779,881,900,930,968,
         1077,1109,1314,1334,1367,1534,1712,1784,1877,1886,2045,
         2056,2260,2429,2509)
```

The median survival time is obtained using median():

```
> median(s)
```

To determine the number of patients with a survival time ≤ 900 days, we perform a logical test and build a table of the individuals based on whether or not they verify the condition:

```
> table(s < 900)
```

The column labelled TRUE indicates the number of observations fulfilling the preceding condition.

Survival associated with the 90th percentile can be obtained by using the command:

```
> quantile(s, 0.9)
```

It can quickly be verified that the result correctly corresponds to the survival value such that 90% of the observations do not exceed this value:

```
> table(s <= quantile(s, 0.9))
```

Here, $38/43$ is really lower than 0.90, whereas $39/43 = 0.91$.

3) The elderly.dat file contains the size measured in cm of 351 elderly females, selected randomly from the population during a study about osteoporosis [EVE 01]. A few observations are nonetheless missing.

a) How many missing observations are there in total?

b) Give a 95% confidence interval for the average size in this Sample, using a normal approximation.

c) Represent the distribution of X as a histogram in the form of a density curve.

To read the file which consists of a series of numeric counts only, the scan() command can be used. Caution should be taken as there are missing frequencies encoded in the text file by "."; it is necessary to ensure that these observations are properly identified as such by R:

```
> sizes <- scan("elderly.dat", na.strings=".")
```

The distribution of observations could be displayed using a simple histogram (histogram(~ sizes)). This allows us to verify the general shape of the distribution and the presence of possible "extreme" frequencies.

The number of incomplete observations (missing data) is determined by counting the number of missing frequencies (NA), which is equivalent to a simple tabulation of the frequencies considered missing by R by means of the table() command:

```
> sum(is.na(sizes))
> table(is.na(sizes))
```

There are, therefore, five missing observations in total.

The mean is obtained using the mean() command. However, as there are missing counts it is necessary to specify the na.rm=TRUE option to instruct R to calculate the arithmetic mean with the complete cases. The same applies to the use of sd(). It is also necessary to have an estimate of the standard error. The reference quantile for the normal distribution (97.5%) is usually taken as 1.96 in statistical tables but its value can be obtained with R using qnorm(), as it was seen above. In the end:

```
> m <- mean(sizes, na.rm=TRUE)      # mean
> s <- sd(sizes, na.rm=TRUE)        # standard deviation
> n <- sum(!is.na(sizes))           # number of observations
> m - qnorm(0.975) * s/sqrt(n)      # 95 % CI low. bound
> m + qnorm(0.975) * s/sqrt(n)      # up. bound. 95 % CI
```

In the presence of missing frequencies, it is important to explicitly ensure that we are working with complete cases. The sum(!is.na(sizes)) command is equivalent to the sum(complete.cases(sizes)) command. The last two instructions can be simplified by making use of the ability of R to repeat the same calculation with data series stored in a variable.

To represent the distribution of sizes as a density curve, which does not raise the problem of the *a priori* choice of classes; the command to use is densityplot():

```
> densityplot(~ sizes)
```

The smoothing degree of the density curve estimated from the data can be controlled using the bw= option. The following example would produce a curve presenting much less local variations, for example:

```
> densityplot(~ sizes, bw=4)
```

3

Measures and Tests of Association Between Two Variables

In this chapter, we will pay particular attention to quantification, in terms of direction and amplitude, and to testing the degree of association between two variables whether symmetrical or not. In the first instance, we will discuss the comparison of mean values, mainly focusing on two samples, whether independent or not, where a variable acts as the response variable and the other as an explanatory variable. The student's t-test will be used, as well as its non-parametric alternative (Wilcoxon–Mann–Whitney test) for two samples. In a second phase, we focus on two-way contingency tables (independent samples) as well as conventional measures of association (chi-squared test) or those more specific to epidemiology (odds ratio, relative risk), including the case of non-independent samples (McNemar's test). In these situations, the variables can play a symmetrical role (or in the case of relative risk). Before describing these testing procedures, the R commands that can be used to summarize a data structure consisting of two variables will also be discussed.

3.1. Bivariate descriptive statistics

3.1.1. *Describing a numeric variable according to the modalities of a qualitative variable*

Consider the variables age of the mother (`age`) and weight of the babies (`low`). The average age of the mothers is to be estimated according to the weight status of the newborn ($< 2,500$ g or within standards). It is perfectly possible to calculate the group averages by selecting the observations belonging to each of the groups or levels (`levels`) of the study factor. For example, in the study on weights at birth the average age of mothers of babies born weighing less than 2.5 kg is equal to:

```
> with(birthwt, mean(age[bwt < 2500]))
[1] 22.30508
```

or equivalently `mean(age[low == "Yes")]`. The procedure would be the same as when calculating the average age of mothers having a child within standards from the point of view of birth weight (`mean(age[low == "No"])`). However, it is often more convenient to group together this kind of operation (calculation of any statistics) by using the `tapply()` command specifying which factor to operate upon:

```
> with(birthwt, tapply(age, low, mean))
      No      Yes
23.66154 22.30508
```

The `tapply()` command relies on the name of the desired response variable upon which the calculations are to be performed being indicated, then the name of the classification variable (usually of the `factor` type) and finally the command to be employed to carry out the calculations. The third argument of the `tapply()` command can be an R command, for example `mean`, `sum`, `summary` or any combination of commands. This strategy is known as "split–apply–combine" [WIC 11] and is illustrated in Figure 3.1: data blocks are constituted for variable y corresponding to the different levels of a factor (variable A); the same command (`mean()`) is applied to each of these blocks and the results are tabulated in the same list.

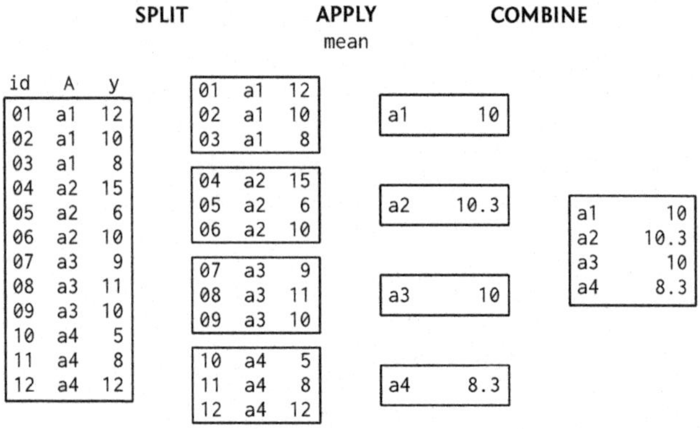

Figure 3.1. *Principle of the "split-apply-combine" approach*

Instead of using R base commands, a "function" could be defined that calculates the mean and standard deviation of a variable. In the following statement, our function is called f and R is instructed to return a list of two values, named m and s, corresponding to the mean and to the deviation of any numeric variable, designated by x, which will be passed to the function f:

```
> f <- function(x) c(m=mean(x), s=sd(x))
> with(birthwt, tapply(age, low, f))
$No
       m        s
23.661538 5.584522

$Yes
       m        s
22.305085 4.511496
```

The distribution of ages according to weight status can be depicted using a histogram, indicating the stratification variable (low) following the | operator. Once again, the formula notation comes forth: here, the objective is to describe the variations in the age variable according to the different levels of low, that is to say conditionally to the variable low. The conditioning is expressed using the symbol |. The advantage of this approach is that R will use the same unit system for the ordinate axis, which often facilitates the comparison between conditional distributions. In addition, the levels of the classification factor are automatically indicated at the top of each chart (Figure 3.2):

```
> histogram(~ age | low, data=birthwt, type="count")
```

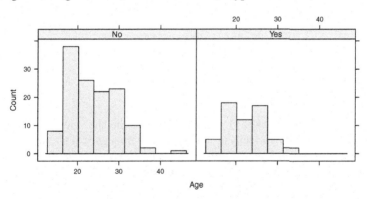

Figure 3.2. *Histogram for the distribution of the age of mothers in relation to the weight status at birth*

Density curves could be used just as well as histograms. The command to be used is then densityplot() and the options are identical to those used with histogram(), with the exception of type="count". In the following illustration, option bw=2.5 has been specified to instruct R to consider a sufficiently large smoothing window that less sensitive to local variations in ages (Figure 3.3):

```
> densityplot(~ age | low, data=birthwt, bw=2.5)
```

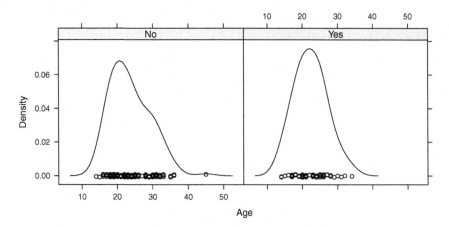

Figure 3.3. *Density curves for the distribution of the age of mothers conditionally to the weight status at birth*

Alternatively we can overlay, or "group", the two density curves in the same chart, as illustrated in Figure 3.4. It is then necessary to add a legend to identify the two sub-samples, and the auto.key=TRUE option provides a legend adapted with automatic placement (by default, in the top margin of the graph):

```
> densityplot(~ age, data=birthwt, groups=low, bw=2.5,
        auto.key=TRUE)
```

3.1.2. Describing two qualitative variables

The table() and xtabs() command, used in the previous chapters to summarize a qualitative variable in the form of a frequency table, can be used to build a table crossing the modalities of two qualitative variables in a contingency table. When looking for the distribution of frequencies by the modalities of the variables low and smoke (weight status in infants and tobacco consumption in mothers), the following will be used:

```
> with(birthwt, table(low, smoke))
    smoke
low   No Yes
  No  86  44
  Yes 29  30
```

Using the xtabs() command, one would write xtabs(~ low + smoke, data= birthwt). Marginal counts are obtained with the margin.table() command by specifying the margin= option (1, totals by rows; 2, totals by columns):

```
> margin.table(xtabs(~ low + smoke, data=birthwt), margin=1)
low
 No Yes
130  59
```

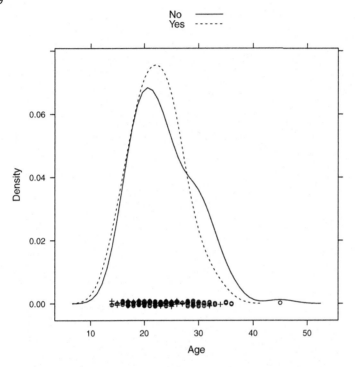

Figure 3.4. *Density curves for the distribution of the age of mothers conditionally to the weight status at birth*

The distribution of babies in the two weight categories (low) can be found with $n = 29 + 30 = 59$ babies having a lower weight than the standard in this sample.

The prop.table() command can always be employed to convert the entries table into a frequency table: by default, R will replace each cell of the contingency table with the value of the count divided by the total count (these are, therefore, conditional frequencies). The margin= option can be used as in the case of margin.table(), to calculate proportions conditionally to one or other of the two variables in the contingency table:

```
> prop.table(xtabs(~ low + smoke, data=birthwt), margin=1)
      smoke
low          No        Yes
```

```
No  0.6615385 0.3384615
Yes 0.4915254 0.5084746
```

This command enables us to determine the frequency of mothers who smoked during pregnancy who had babies with a lower-than-average weight ($44/(44 + 86) =$ 34%).

Similarly to the univariate case, it is possible to represent the distribution of the frequencies for two qualitative variables using either a bar chart (barchart() or a dot plot (or Cleveland's plot, dotplot()). In both cases, these commands expect a frequency table and not a list of variables related by a formula notation (except in the case where the table is turned into a three-column data frame in which the levels of each variable and the counts associated with their crossing are explicitly revealed, for example as.data.frame(xtabs(~ low+smoke, data=birthwt))). For example, with variables low and smoke in a bar chart (Figure 3.5):

```
> barchart(xtabs(~ low + smoke, data=birthwt), stack=FALSE,
      auto.key=TRUE, ylab="low")
```

The stack=FALSE option makes it possible to display separate bars for the modalities of the low variable, recalled in the legend by using auto.key=TRUE. The orientation of the bars is controlled with the horizontal= option (by default, it is equal to TRUE).

3.2. Comparisons of two group means

"Simple" test procedures are generally accessible through R commands including the keyword test: t.test() (student t-test), chisq.test() (Pearson's χ^2 test), etc. In some cases, they are based on an R formula notation according to which a response variable "described" by an explanatory variable is designated; for example, y ~ x generally designates the response y described by variable x (of the factor type).

3.2.1. Independent samples

In the case of two independent samples, the student's t-test for testing the equality of group means is obtained with the t.test() command. To compare the mean weight of mothers having presented with interuterine pain (variable ui) and that of mothers with no problems, the following instructions will be executed:

```
> t.test(lwt ~ ui, data=birthwt, var.equal=TRUE)

Two Sample t-test

data:  lwt by ui
t = 2.1138, df = 187, p-value = 0.03586
alt hypothesis: true difference in means is not equal to 0
95 percent confidence interval:
  0.8753389 25.3544748
sample estimates:
 mean in group No mean in group Yes
         131.7578          118.6429
```

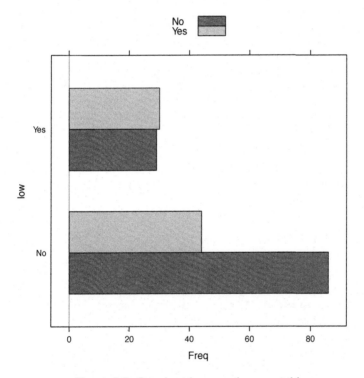

Figure 3.5. *Bar chart for a contingency table*

The formula notation enables R to be instructed that the relationship between the two variables lwt and ui is the main focus, where the variations of lwt are "described" by ui. In other words, lwt is considered a response variable whose average variations have to be explained with regard to classification variable ui. The var.equal=TRUE option (by default, it is equal to FALSE) instructs R not to apply a

Satterwaithe-type correction (so-called Welch test) during the calculation. R displays the result of the null hypothesis test by giving the value of the test statistic (t), the number of degrees of freedom (df) and the associated degree of significance (p-value). It also provides a confidence interval for the parameter of interest, here the difference between the means.

Naturally, the conditions for the application of this test can be verified, the equality of the population variances, by comparing the values of the variances estimated from the sample:

```
> with(birthwt, tapply(lwt, ui, var))
       No      Yes
940.8472 783.7196
```

An alternative consists of visually verifying the distribution of lwt in both groups defined by variable ui, for example by using a boxplot-type diagram (Figure 3.6):

```
> bwplot(lwt ~ ui, data = birthwt, pch = "|")
```

A more formal test for the comparison of the two variances is available through the var.test() command. The previous results can also be compared with those that would be obtained by correcting the degrees of freedom of the reference distribution t to take into account the heterogeneity of the variances (the degrees of freedom are generally no longer integers). To achieve this, we can remove the var.equal=TRUE option when using t.test():

```
> t.test(lwt ~ ui, data=birthwt)

Welch Two Sample t-test

data:  lwt by ui
t = 2.2547, df = 39.163, p-value = 0.02982
alt hypothesis: true difference in means is not equal to 0
95 percent confidence interval:
  1.351128 24.878685
sample estimates:
 mean in group No mean in group Yes
         131.7578          118.6429
```

The tests previously performed consider a possible non-oriented alternative (difference of the two non-zero means, but without specifying the direction of the difference) and are, therefore, two-tailed. To specify the direction of the difference expected under the alternative hypothesis (for example, $H_1 : \mu_1 > \mu_2$), the alternative= option must be entered in the t.test() command.

variable

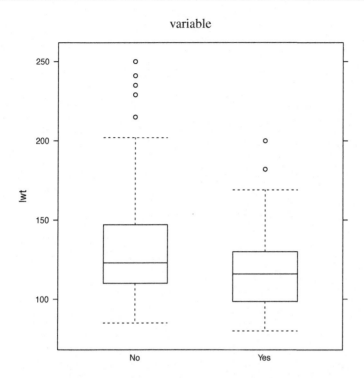

Figure 3.6. *Distribution of the lwt variable according to the levels of the ui variable*

3.2.2. *Non-independent samples*

In the case where the samples are not independent, the R command remains `t.test()` but the `paired=TRUE` option must be added to instruct R that the two samples are not independent (e.g. cases where the two series of measurements were collected from the same statistical units on two occasions).

This section will discuss the possible organization of the data in a table while working with two sets of paired measures. Consider two sets of scores on a depression scale (Hamilton scale) obtained from nine patients before (x1) and after treatment (x2) using tranquilizers [HOL 73]:

```
> x1 <- c(1.83,0.50,1.62,2.48,1.68,1.88,1.55,3.06,1.30)
> x2 <- c(0.878,0.647,0.598,2.05,1.06,1.29,1.06,3.14,1.29)
```

The natural way to organize these data would consist of arranging these two series of measures in a table with nine rows and two columns, the columns representing the two previous variables:

```
> d <- data.frame(x1, x2)
> dim(d)
[1] 9 2
> summary(d)
       x1                x2
 Min.   :0.500    Min.   :0.598
 1st Qu.:1.550    1st Qu.:0.878
 Median :1.680    Median :1.060
 Mean   :1.767    Mean   :1.335
 3rd Qu.:1.880    3rd Qu.:1.290
 Max.   :3.060    Max.   :3.140
```

A t test to compare the means of the two periods can then be described as follows:

```
> with(d, t.test(x1, x2, paired=TRUE))

Paired t-test

data:  x1 and x2
t = 3.0354, df = 8, p-value = 0.01618
alternative hypothesis: true difference in means is not
  equal to 0
95 percent confidence interval:
 0.1037787 0.7599991
sample estimates:
mean of the differences
              0.4318889
```

Instead one could place the depression scores in one column and the measurement period (before or after treatment) in another. To transform the table d (two columns, x1 and x2) into a table containing these two new variables, the melt() command of the reshape2 package can be used. It will be necessary to install this with the command install.packages() in R. The library() command is then used to "load" the package inside the R session and to get access to the commands it provides:

```
> library(reshape2)
> dm <- melt(d)
No id variables; using all as measure variables
> dim(dm)
[1] 18  2
> summary(dm)
 variable      value
```

```
x1:9      Min.   :0.500
x2:9      1st Qu.:1.060
          Median :1.425
          Mean   :1.551
          3rd Qu.:1.867
          Max.   :3.140
```

It can be verified that the same means can be found correctly in each group with the summary(d) command used above:

```
> with(dm, tapply(value, variable, mean))
      x1        x2
1.766667 1.334778
```

In this instance the t test would be now written as t.test(value~variable, data=dm, paired= TRUE).

Note that it would also be possible to make more explicit the fact that these are the same patients measured in both experimental conditions (x1 and x2) by adding a variable encoding the identifier of the patient (number):

```
> dm$id <- factor(rep(1:9, 2))
> head(dm, n=3)
  variable value id
1       x1  1.83  1
2       x1  0.50  2
3       x1  1.62  3
```

To represent the individual data graphically while taking account of the pairing, two graphical solutions are available:

```
> library(gridExtra)
> p1 <- stripplot(value ~ variable, data=dm, groups=id, type="l",
              col="grey50", lwd=1.2)
> p2 <- xyplot(x2 ~ x1, data=d, type=c("p", "g"),
            abline=list(a=0, b=1))
> grid.arrange(p1, p2, nrow=1)
```

The gridExtra package (to be installed) provides the grid.arrange() command that allows different graphic to be assembled on a single graphic window. By examining Figure 3.7(b), it can be verified that the majority of the points are located below the diagonal, represented by the straight line with unit slope. This suggests that patients generally have a lower score after treatment (x2). In Figure 3.7(a), this translates into individual negative slopes in most cases.

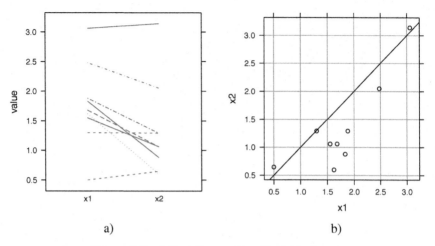

a)   b)

Figure 3.7. *a) Evolution profile of individual scores;*
b) scatter plot of the scores

3.3. Comparisons of proportions

3.3.1. *Case of two proportions*

Suppose that one wants to consider the probability of observing under-weight children at birth (low) according to whether their mothers smoked or not during their pregnancy (smoke). The null hypothesis is that these two probabilities are equal. The prop.test() command allows us to test this hypothesis using the asymptotic normal distribution:

```
> tab <- with(birthwt, table(smoke, low))
> tab
      low
smoke No Yes
   No  86  29
   Yes 44  30
> prop.table(tab, margin=1)
      low
smoke        No       Yes
   No  0.7478261 0.2521739
   Yes 0.5945946 0.4054054
> prop.test(tab[,c(2,1)], correct=FALSE)

2-sample test for equality of proportions without continuity
correction
```

```
data:  tab[, c(2, 1)]
X-squared = 4.9237, df = 1, p-value = 0.02649
alternative hypothesis: two.sided
95 percent confidence interval:
 -0.29039121 -0.01607177
sample estimates:
   prop 1     prop 2
0.2521739 0.4054054
```

Care must be taken as the table must be reorganized slightly for the event of interest (low="Yes") to appear in the first column, hence the permutation of the two columns of the table with tab[,c(2,1)] and the desired proportions 30/74 (smoking) and 29/115 (non-smoking). This does not alter the result of the hypothesis test but it obviously affects the estimated proportions and the confidence interval associated with the difference between these two proportions.

An alternative formulation consists of indicating the number of "successes" (here, the fact that the baby be under-weight) and the total frequency for each class defined by the smoke variable:

```
> prop.test(c(29,30), c(86+29, 44+30), correct=FALSE)

2-sample test for equality of proportions without continuity
correction
data:  c(29, 30) out of c(86 + 29, 44 + 30)
X-squared = 4.9237, df = 1, p-value = 0.02649
alternative hypothesis: two.sided
95 percent confidence interval:
 -0.29039121 -0.01607177
sample estimates:
   prop 1     prop 2
0.2521739 0.4054054
```

The prop.test() command includes a correct= option that enables the decision whether a continuity correction (Yates correction) needs to be applied or not.

3.3.2. Chi-squared test

Another test frequently used while working with contingency tables is Pearson's χ^2 test, which aims to test the association between two qualitative variables, the deviation from the independence being quantified by the difference between the counts expected under the independence hypothesis and the counts observed from the sample. An application considering the same low and smoke variables follows:

```
> chisq.test(xtabs(~ low + smoke, data=birthwt))
```

Pearson's Chi-squared test with Yates' continuity correction

```
data:  xtabs(~low + smoke, data = birthwt)
X-squared = 4.2359, df = 1, p-value = 0.03958
```

It should be noted that the previous statement is equivalent to:

```
> with(birthwt, chisq.test(table(low, smoke)))
> summary(xtabs(~ low + smoke, data=birthwt))
```

As in the case of prop.test(), the correct= option can be used in the case of a 2×2 table to achieve a continuity correction.

By storing the previous result in a dummy variable, it is possible to verify whether the conditions for the application of this test are verified, that is to say the theoretical values (usually > 5) are large enough:

```
> res <- chisq.test(xtabs(~ low + smoke, data=birthwt))
> res$expected
        smoke
low         No       Yes
   No  79.10053 50.89947
   Yes 35.89947 23.10053
```

As can be observed, the addition $expected following the variable in which the result of the χ^2 test has been stored is sufficient. In fact, the statistical tests achieved by R return numerous values (test statistic, degrees of freedom, etc.) and R automatically displays the relevant information to make a decision concerning the result of the test. However, other information remains hidden but can be found following the same procedure as illustrated previously. In the case of the χ^2 test, the str(res) command makes it possible to display the information that is available (observed, expected counts and residuals, among others).

3.3.3. The case of non-independent samples

In the case where the contingency table is formed from the crossing of two variables observed in the same sample (for example, severity of the two-class symptomatology in pre and post treatment), one can resort to McNemar's test using the mcnemar.test() command. This command also accepts the data presented in the form of a table generated with `table()` or `xtabs()`. The online help provides an example about how to use this command, see `help(mcnemar.test)`.

3.4. Risk and odds ratio measures

The risk measures usually calculated from a contingency table crossing an exposure factor (or any binary variable of interest) and a binary response variable (sick/well, failure/success) are the relative risk and the odds ratio. The latter can be used more generally to quantify the degree of association between two binary variables.

The data are usually arranged in a 2×2 contingency table where the frequencies related to the "positive events" appear in the first row and in the first column. The variables `smoke` and `low` used in the preceding illustrations have two levels, 0 and 1, and by default the reference level 0 (encoding for negative events, non-smoker or normal weight) are the ones that will appear in the first row and in the first column when using `table()` or `xtabs()`. This remains valid when using the labels No (0) and Yes (1). In practice, this does not raise any problem for the calculation of the odds ratio, but care needs to be taken for the relative risk. The contingency table will thus be rearranged by swapping the two rows and the two columns:

```
> tab <- xtabs(~ smoke + low, data=birthwt)
> tab <- tab[c(2,1),c(2,1)]
> tab
       low
smoke Yes No
   Yes  30 44
   No   29 86
```

It thus can be seen that the first row of the `tab` table contains the distribution of children having a weight lower or greater than 2.5 kg for smoking mothers and that the first column contains the distribution of smoking or non-smoking mothers for underweight children at birth. The odds ratio is simply the ratio of odds 30/44 and 29/86, or $\frac{30/44}{29/86} = \frac{30 \times 86}{44 \times 29} = 2.02$. The relative risk represents the risk ratio of each exposure group. Here, the fact that the mother was smoking during her pregnancy represents the exposure factor, hence a risk of $30/(30 + 44)$ in smokers compared to $29/(29 + 86)$ in non-smokers, that is a relative risk of $\frac{30/(30+44)}{29/(29+86)} = 1.61$.

There are many commands that make it possible to repeat these manual calculations and to provide confidence intervals for the estimated values of the odds ratio or of the relative risk. For the calculation of the odds ratio, an external package will be used, vcd, that will have to be installed. The `install.packages("vcd")` command will be entered in R and after verifying that the installation is successfully completed, and the package will be loaded by using the `library()` command:

```
> library(vcd)
> oddsratio(tab, log=FALSE)
[1] 2.021944
```

The `log=FALSE` option is essential if the value of the odds ratio has to be obtained and not its logarithm. The `confint()` command enables us to obtain the associated confidence interval, by default at 95%: `confint(oddsratio(tab, log=FALSE))`.

Alternatively, we can use the epiR package, which also needs to be installed and loaded before the commands provided in this package can be used. The advantage of this package and of the `epi.2by2()` command is that they provide an estimate of the relative risk and odds ratio, considering a cohort study for which these two indicators are meaningful by default:

```
> library(epiR)
> epi.2by2(tab)
            Outcome +   Outcome -   Total   Inc risk *    Odds
Exposed +        30          44      74        40.5      0.682
Exposed -        29          86     115        25.2      0.337
Total            59         130     189        31.2      0.454

Point estimates and 95 % CIs:
-------------------------------------------------------------------
Inc risk ratio                            1.61 (1.06, 2.44)
Odds ratio                                2.01 (1.03, 3.96)
Attrib risk *                            15.32 (1.61, 29.04)
Attrib risk in population *               6.00 (-4.33, 16.33)
Attrib fraction in exposed (%)           37.80 (5.47, 59.07)
Attrib fraction in population (%)        19.22 (-0.21, 34.88)
-------------------------------------------------------------------
 * Cases per 100 population units
```

3.5. Non-parametric approaches and exact tests

For the comparison of numerical measurements collected over two independent samples, a non-parametric approach based solely on the ranks of the observations consists of using the Wilcoxon–Mann–Whitney test, available in the `wilcox.test()` command. As in the case of the student's t-test, when the samples are not independent, the paired=TRUE option will be added. With the two examples previously discussed (section 3.2), one will proceed in the following manner:

```
> wilcox.test(lwt ~ ui, data=birthwt)

Wilcoxon rank sum test with continuity correction

data:  lwt by ui
W = 2896, p-value = 0.01626
alternative hypothesis: true location shift is not equal to 0

> with(d, wilcox.test(x1, x2, paired=TRUE))

Wilcoxon signed rank test

data:  x1 and x2
V = 40, p-value = 0.03906
alternative hypothesis: true location shift is not equal to 0
```

The second test for the case of paired samples is called the signed-rank test. With this example using variables x1 and x2 in the data frame d, an equivalent formulation using the table showing distinctly explanatory variables and response variables (dm) would be:

```
> wilcox.test(value ~ variable, data=dm, paired=TRUE)
```

With respect to contingency tables, an alternative to the χ^2 test is the Fisher's exact test available through the fisher.test() command:

```
> fisher.test(xtabs(~ low + smoke, data=birthwt))

Fisher's Exact Test for Count Data

data:  xtabs(~low + smoke, data = birthwt)
p-value = 0.03618
alternative hypothesis: true odds ratio is not equal to 1
95 percent confidence interval:
 1.028780 3.964904
sample estimates:
odds ratio
  2.014137
```

It should be noted that this command also returns the odds ratio and its 95% confidence interval.

Finally, R includes the `binom.test()` command to perform a binomial test. It allows for testing the probability of success in the case of a binary variable assuming a binomial distribution and not a normal distribution, as in the case of `prop.test()`.

3.6. Key points

– Before performing a test, a few descriptive statistics are always computed for each level of an explanatory variable (`tapply()`) or by crossing the modalities of two qualitative variables (`table()` or `xtabs()`).

– The graphical representations of two variables do not differ from those met with the univariate case, with the exception that a formula notation is being used to indicate the relationships between variables. In the case of qualitative variables, a table of frequencies is typically provided (`barchart()` or `dotplot()`).

– The test procedures usually return the result of the null hypothesis associated with the test and a confidence interval for the parameter estimated.

– In the case of risk measures (odds ratio, but especially relative risk), caution should be taken when arranging the counts (in rows and in columns) in a contingency table.

3.7. Going further

The `psych` package (www.personality-project.org/r/psych/) provides a few commands specific for the calculation and the graphical representation of conditional means (`describeBy()` and `error.bars.by()`). The `plyr` (http://plyr.had.co.nz) and `dplyr` packages also offer the possibility of calculating statistics with aggregated data in a relatively effective way. Similar functionality is available in the `Hmisc` package, described in Appendix 3.

In addition to the `power.t.test()` and `power.prop.test()` commands, the `pwr` and `clinfun` packages include commands specific to the calculation of the number of subjects needed to give emphasis to a mean or to a proportional difference defined *a priori* for a given power and a given alpha risk.

3.8. Applications

1) The quality of sleep in ten patients was measured before (control) and after treatment with one of the two following hypnotics: (1) D. hyoscyamine hydrobromide and (2) L. hyoscyamine hydrobromide. The assessment criterion retained by the researchers was the average gain of sleep (in hours) compared to the basic duration of sleep (control) [STU 08]. The data are reported below and are also part of the basic datasets (`data(sleep)`) of R.

The researchers concluded that only the second molecule actually had a soporific effect.

a) Estimate the average time of sleep for each of the two molecules, as well as the difference between these two averages.

b) Display the distribution of difference scores (LHH – DHH) in the form of a histogram, considering half-hour class intervals, and indicate the mean and the standard deviation of the difference scores.

c) Verify the accuracy of the findings using a student's t-test.

The same patients are being tested in both conditions (the subject is taken as its own witness). The online help for the sleep dataset indicates that the two first variables, extra (sleep time differential) and group (type of medicine), are the variables of interest:

```
> data(sleep)
> mean(sleep$extra[sleep$group == 1]) # D. hyoscyamine hydro.
> mean(sleep$extra[sleep$group == 2]) # L. hyoscyamine hydro.
> m <- with(sleep, tapply(extra, group, mean))
> m[2] - m[1]
```

The previous commands allow verification of the group means and the difference of gain in sleep between the *L* hyoscyamine hydrobromide and *D* hyoscyamine hydrobromide molecules, keeping in mind the with(sleep, tapply (extra, group, mean)) command returns two values (the means by type of molecule) and the difference between these two values (stored in the auxiliary variable m) properly returns the mean difference between the two treatments.

To calculate the difference scores, one can proceed in the following manner:

```
> sdif <- sleep$extra[sleep$group == 2] -
          sleep$extra[sleep$group == 1]
> c(mean=mean(sdif), sd=sd(sdif))
```

The calculation of the mean and standard deviation of the difference scores raises no particular difficulty, nor does displaying their distribution in the form of a histogram:

```
> histogram(~ sdif, breaks=seq(0, 5, by=0.5), xlab="LHH - DHH")
```

The `breaks=seq(0, 5, by=0.5)` option makes it possible to build a sequence of numbers between 0 and 5 and separated by 0.5 units for the break down of the classes measured.

The result of the t-test is obtained with the `t.test()` command providing the response variable and the classification factor in the form of a formula (response variable on the left, explanatory variable on the right):

```
> t.test(extra ~ group, data=sleep, paired=TRUE)
```

It should be noted that it is necessary to instruct R where to find the variables of interest, hence, the use of `data=sleep`. The `paired=TRUE` option is required to account for the pairing of the observations. Observe that, by default, R presents the difference in means between the first and the second treatment and not the second minus the first, as calculated previously.

The significant result of the test and the meaning of the mean difference being observed (at the level of the gain in sleep time) is effectively in agreement with the conclusions of researchers. The results can be visualized in the form of a bar chart (a dot plot could also used by means of the `dotplot()`) command:

```
> barchart(with(sleep, tapply(extra, group, mean)))
```

2) In a clinical trial, the objective was to evaluate a system supposed to reduce the number of symptoms associated with benign breast diseases. A group of 229 women with these diseases were randomly divided into two groups. The first group received routine care, while the patients in the second group followed a special regimen (variable B = treatment). After one year, individuals were evaluated and classified in one of two categories: improvement or no improvement (variable A = response) [SEL 98]. The results are summarized in Table 3.1, for a portion of the sample.

a) Perform a chi-squared test.

b) What are the theoretical frequencies expected under an independence hypothesis?

c) Compare the results obtained in (a) with those of a Fisher's exact test.

d) Provide a 95% confidence interval for the proportion improvement difference between the two groups of patients.

Only the data regarding the measures presented in the table are available, but it is not necessary to access the raw data to perform the χ^2 test. This is obtained with the chisq.test() command and by default includes a continuity correction (Yates). To build the contingency table, the matrix() command can be used:

```
> regime <- matrix(c(26,38,21,44), nrow=2)
> dimnames(diet)<-list(c("improvement","no improvement"),
                       c("diet","no diet"))
```

	Diet	No diet	Total
Improvement	26	21	47
No improvement	38	44	82
Total	64	65	129

Table 3.1. *Diet and breast disease*

By default, the measures are captured by column, unless the byrow=TRUE option is specified. In the case of a 2 x 2 table, it is not necessary to specify the number of rows when the number of columns is indicated (or vice versa). The χ^2 test is obtained using chisq.test():

```
> chisq.test(diet)
```

If no continuity correction is to be applied, it is necessary to add the correct=FALSE option.

Theoretical frequencies are not displayed with the result of the test, but they can be obtained as follows:

```
> chisq.test(diet)$expected
```

With regards to Fisher's exact test and following the same procedure, the command fisher.test() is employed from the contingency table:

```
> fisher.test(diet)
```

The proportions of interest are $26/64 = 0.41$ (diet) and $21/65 = 0.32$ (no diet), hence the proportion test that gives the 95% confidence interval for the desired difference:

```
> prop.test(c(26,21), c(64,65))
```

3) A study focused on 86 children followed learning how to manage their disease. On arrival at the center, the children were asked if they knew how to manage their disease in optimal conditions (not detailed), that is if they knew when they had to resort to the prescribed treatment. They were asked the same question at the end of their follow-up at the center. The variable measured is the answer ("yes" or "no") to this question at the start and at the end of the study [PEA 05]. The data, available in SPSS format in the health-camp.sav file, are summarized in Table 3.2.

			Disease management (discharge)		
			No	Yes	Total
	No	Counts	27	29	56
		Proportion	31.4%	33.7%	65.1%
Disease management	Yes	Counts	6	24	30
(admission)		Proportion	7.0%	27.9%	34.9%
	Total	Counts	33	53	86
		Proportion	38.4%	61.6%	100.0%

Table 3.2. *Disease management*

It can be argued whether having followed the training program proposed at the center increased the number of children with a good knowledge of their illness and of its daily management.

a) Reproduce the previous measures and relative frequencies table from the raw data.

b) Indicate the result of a McNemar's test.

c) Compare the result of the test performed in (b) but without correction continuity with the result obtained from a binomial test.

It is not really necessary to import the raw data table and one could merely work with the frequencies presented in the table. However, to load the SPSS file, a specific R command, available in the foreign package has to be used:

```
> library(foreign)
> hc <- read.spss("health-camp.sav", to.data.frame=TRUE)
```

The to.data.frame=TRUE option is important because it is the one that ensures that after importing them, the data will be stored correctly in a table where the rows present the observations, and the columns the variables. If we verify how data are represented in the table imported using the head() command, it will be concluded that the information is slightly vague:

```
> head(hc)
```

This is due to the loss of the labels associated with the variables in SPSS. This information can be found by querying the R database using the `str()` command.

```
> str(hc)
```

Starting from the `attr(*, "variable.labels")` row, the BEFORE and AFTER columns correspond to the question regarding the knowledge of the disease, whereas the BEFORE2 and AFTER2 columns correspond to the question regarding the treatment.

Finally, the original table (measures and relative frequencies) can be reproduced as follows:

```
> table(hc[,c("BEFORE","AFTER")])
> round(prop.table(table(hc[,c("BEFORE","AFTER")])), 2)
```

To obtain the marginal distributions, the `margin.table()` command can be used:

```
> margin.table(table(hc[,c("BEFORE","AFTER")]), 1)
> margin.table(table(hc[,c("BEFORE","AFTER")]), 2)
```

To perform McNemar's test, one can use the contingency table previously built. For the sake of convenience, the latter will be stored in an R variable called hc.tab:

```
> hc.tab <- table(hc[,c("BEFORE","AFTER")])
> mcnemar.test(hc.tab)
```

McNemar's test results can be compared without applying the continuity correction with those of a binomial test (exact test):

```
> binom.test(6, 6+29)
> mcnemar.test(hc.tab, correct=FALSE)
```

4

Analysis of Variance
and Experimental Design

In this chapter, we will address analysis of variance and a few elementary notions about the design of experiments. Only the ANOVA cases with one and two classification factors (fixed effects, interaction model) are discussed. For a single-factor ANOVA, the linear tendency test is approached using two techniques: linear regression and the contrast method. The case of multiple comparisons between treatments is also discussed, considering only the Bonferroni method to protect the inflation of the Type I error risk. It remains, however, important to achieve an in-depth data description in all cases prior to performing the ANOVA model (numerical summaries, graphics methods, etc.).

4.1. Data representation and descriptive statistics

4.1.1. *Data representation format*

When considering a numerical variable acting as a response variable and a k-level qualitative variable (modalities or treatments) used as classification factor, at least two ways to represent the data in R can be devised: a table where the k series of observations are represented in different columns (a table with k columns and n rows in the case of a balanced design where each treatment has the same number of observations) or a table where all the measures appear in a first column and the associated treatments in a second column, as discussed in section 3.2.2

Caution should still be taken due to the fact that the classification factor must effectively be considered as a `factor` and not as a numeric variable.

4.1.2. *Descriptive statistics and data structuring*

In the following example a data structure is generated, composed of twenty observations distributed in a balanced manner over the four levels (treatments) of a qualitative variable. The data are arranged in four columns, each column corresponding to a specific treatment, resulting in a table with five rows and four columns. The measures are artificially generated from sampling without replacement (replicate) in a normal distribution with mean 12 and standard deviation 1.2:

```
> d <- data.frame(replicate(4, rnorm(5, mean=12, sd=1.2)))
> d
        X1        X2        X3        X4
1 12.67078  9.729374 12.47485 12.38534
2 12.52286 11.428080 10.16006 11.86140
3 12.77444 11.866122 12.42084 13.01769
4 10.20481 10.663182 10.90229 10.45179
5 13.55448 11.733637 10.49526 10.76963
> summary(d)
       X1              X2              X3              X4
 Min.   :10.20   Min.   : 9.729   Min.   :10.16   Min.   :10.45
 1st Qu.:12.52   1st Qu.:10.663   1st Qu.:10.50   1st Qu.:10.77
 Median :12.67   Median :11.428   Median :10.90   Median :11.86
 Mean   :12.35   Mean   :11.084   Mean   :11.29   Mean   :11.70
 3rd Qu.:12.77   3rd Qu.:11.734   3rd Qu.:12.42   3rd Qu.:12.39
 Max.   :13.55   Max.   :11.866   Max.   :12.47   Max.   :13.02
```

The average of the measurements has to be to calculated for each column. In the case where the data table is a data frame, the simplest way to apply the same command to each column of the table consists of using the command sapply() as follows:

```
> sapply(d, mean)
      X1       X2       X3       X4
12.34548 11.08408 11.29066 11.69717
```

Suppose now that the same data be arranged in a table with two columns, the first column indicating the treatment and the second column the captured measurement. For the sake of simplicity, the observation number will not appear in another column. The package reshape2 (to previously load) provides the command melt() which has already been used and that allows a table in which the levels of a factor are arranged in columns to be transformed into a table where these are arranged in the same column:

```
> dm <- melt(d, value.name="y", variable.name="trt")
No id variables; using all as measure variables
```

```
> head(dm)
   trt        y
1  X1  12.670780
2  X1  12.522859
3  X1  12.774442
4  X1  10.204814
5  X1  13.554482
6  X2   9.729374
> summary(dm)
   trt          y
X1:5    Min.    : 9.729
X2:5    1st Qu.:10.621
X3:5    Median :11.798
X4:5    Mean    :11.604
        3rd Qu.:12.487
        Max.    :13.554
> with(dm, tapply(y, trt, mean))
      X1       X2       X3       X4
12.34548 11.08408 11.29066 11.69717
```

It can be seen that with this type of structure, we are once again confronted with the situations mentioned in Chapters 2 and 3 and that the command tapply() can be used without any problem. As a general rule, this type of data organization for statistical models will be the most preferable to achieve in R.

A command similar to tapply() but that enables the usage of formula notation is the command aggregate() whose utilization is relatively simple: indications are provided about the type of relationship of interest between two (or more) variables, as well as about the data frame where to find the variables and the command to apply, here mean():

```
> aggregate(y ~ trt, data=dm, mean)
   trt        y
1  X1  12.34548
2  X2  11.08408
3  X3  11.29066
4  X4  11.69717
```

This command is strictly equivalent to the previous instruction tapply(), only the format of the returned result is different: the command aggregate() returns its results in the form of a data frame which sometimes can be useful for secondary analyses or for graphic representations.

4.2. One-way ANOVA

4.2.1. The one-way ANOVA model

Consider the example of the variations between the weights of babies (bwt) according to the ethnicity of the mother (race) in the data birthwt. The question that may be asked is: does the average weight of babies at birth differ according to the ethnicity of the mother?

The ANOVA model is built using the command aov() in R, indicating by means of a formula the relationship between the response variable (on the left of the symbol ~) and the explanatory variable (to the right of the symbol ~):

```
> aov(bwt ~ race, data=birthwt)
Call:
   aov(formula = bwt ~ race, data = birthwt)

Terms:
                    race  Residuals
Sum of Squares   5015725   94953931
Deg. of Freedom        2        186

Residual standard error: 714.4963
Estimated effects may be unbalanced
```

It can be seen that the result returned by R does not really resemble an ANOVA table and that it contains sums of squares only (with their degrees of freedom) and an estimation of the residual variance (sqrt(96179472/187)). To view the ANOVA table corresponding to the model bwt ~ race, the command summary() must be used. This is the same command as that utilized to summarize a data frame or a variable but it behaves differently in the case where the R object in question is an ANOVA model built with the command aov() (more specifically, the command is actually summary.aov):

```
> summary(aov(bwt ~ race, data=birthwt))
             Df   Sum Sq  Mean Sq  F value  Pr(>F)
race          2  5015725  2507863    4.913  0.00834 **
Residuals   186 94953931   510505
---
Signif. codes:  0 '***' 0.001 '**' 0.01 '*' 0.05 '.' 0.1 ' ' 1
```

It should be noted that it will often be more convenient to store the results of the ANOVA model in a variable for later reuse with other commands. The two previous steps can, therefore, be reformulated as follows:

```
> res <- aov(bwt ~ race, data=birthwt)
> summary(res)
```

A natural graphic representation for this type of data (a numeric variable described by a qualitative variable) is a box plot type of diagram, to gain an idea of fluctuations within groups or treatments and a bar or point diagram to get an idea of the variation of the means by treatment:

```
> bwplot(bwt ~ race, data=birthwt)
```

It should be observed that the formula notation is identical to that used to construct the ANOVA model. With regard to the dot plot, a small variation should be introduced:

```
> dotplot(race ~ bwt, data=birthwt, type="a")
```

The option `type="a"` enables the automatic calculation of the averages of bwt per variable level `race`. This latter is pictured here on the left of the symbol ~ in the R formula which makes it possible to represent it on the y-axis rather than on the x-axis. The result produced by the commands `bwplot()` and `dotplot()` from this data set is illustrated in Figure 4.1. In the case of box plots, the points inside the boxes represent the medians in each group. For the bar chart, the group averages are connected by segments to facilitate reading.

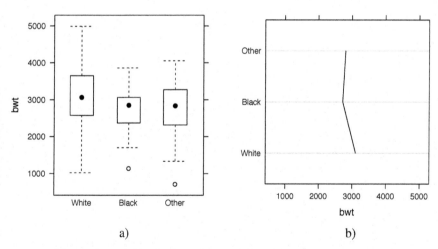

a)   b)

Figure 4.1. *a) Distribution of the weights of babies according to the ethnicity of the mother using box plots; b) average weights of babies according to ethnicity*

An equivalent way to summarize the data observed is to use a table of group averages directly, using `aggregate()` presented in the previous section. Here follows an illustration replacing dot plot (dotplot()) by a bar chart using barchart() (Figure 4.2):

```
> bwtmeans <- aggregate(bwt ~ race, data=birthwt, mean)
> bwtmeans
    race      bwt
1 White 3102.719
2 Black 2719.692
3 Other 2805.284
> barchart(bwt ~ race, data=bwtmeans, horizontal=FALSE,
          ylim=c(2350, 3250))
```

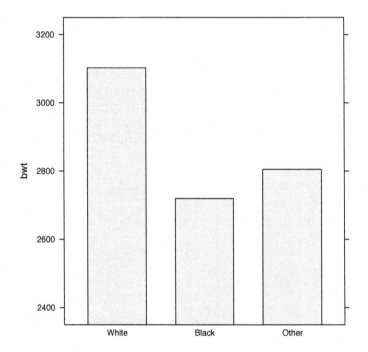

Figure 4.2. *Distribution of the weights of babies according to the ethnicity of the mother using a bar chart*

It would be naturally possible to add the deviations or confidence intervals associated with these group averages.

To verify the conditions for the application of the ANOVA (normality of residuals and homogeneity of variances), graphical representations or formal tests can be used. The Shapiro-Wilk test (distribution normality, case of large samples) is accessible through the command shapiro.test(), but graphical methods (QQ plot or box plots) are generally preferred. With regard to the equality of variances, Levene (command leveneTest() in the package car) or Bartlett (bartlett.test ()) tests are available. These commands use the same formula notation:

```
> bartlett.test(bwt ~ race, data=birthwt)

Bartlett test of homogeneity of variances

data:  bwt by race
Bartlett's K-squared = 0.65952, df = 2, p-value = 0.7191
```

Finally, the commands fitted() (or predict) and resid() allow us to obtain the values predicted by the model and residuals (deviations between observed and predicted values). It shall be verified without difficulty that in this case the predicted values correspond to the averages of the treatments:

```
> unique(predict(res))
[1] 2719.692 2805.284 3102.719
```

4.2.2. Comparisons using pairs of treatments

A significant result for the Fisher-Snedecor F-test in ANOVA signifies that at least a pair of means can be considered statistically different at the error threshold priorly defined (5% as a general rule). Several so-called "multiple comparisons" techniques have been proposed in the literature. Only the Bonferroni method will be presented here for the comparison of all the pairs of averages using Student's t-tests.

A fairly conservative approach involves applying a Bonferroni correction to the tests carried out: this is equivalent to multiplying each level of significance by the total number of tests performed. The command pairwise.t.test() makes it possible to test all pairs of averages using Student's t-tests and returns the levels of significance corrected for each pair of treatments:

```
> with(birthwt, pairwise.t.test(bwt, race,
                              p.adjust.method="bonferroni"))

Pairwise comparisons using t tests with pooled SD

data:  bwt and race

      White Black
Black 0.049 -
```

```
Other 0.029 1.000
```

```
P value adjustment method: bonferroni
```

It can be verified that the same results are obtained properly by means of simple Student's t-tests, using t.test() and excluding systematically one of the levels of the variable race.

Here follows an example with the comparisons White/Black and White/Other:

```
> t1 <- t.test(bwt ~ race, data=birthwt, subset=race != "Other")
> t2 <- t.test(bwt ~ race, data=birthwt, subset=race != "Black")
> c(t1$p.value, t2$p.value) * 3
[1] 0.03509656 0.03277981
> with(birthwt, pairwise.t.test(bwt, race, p.adj="bonf",
                         pool.sd=FALSE))
```

```
Pairwise comparisons using t tests with non-pooled SD
```

```
data:  bwt and race
       White Black
Black 0.035 -
Other 0.033 1.000
```

```
P value adjustment method: bonferroni
```

For the comparison between the two approaches to be valid, it is necessary to instruct R to not use an estimate of the variance common to all of the treatments (as in ANOVA) but that only of the treatments taken into account in the comparison, hence, the option pool.sd=FALSE. Finally, it shall be retained that when the options are not misleading, they may be abbreviated; in the above case, it is perfectly possible to write p.adj="bonf" instead of p.adjust.method = "bonferroni".

4.2.3. Linear trend test

In the cases where it makes sense to consider that the levels of factor classification are ordered, it may be interesting to formulate the ANOVA in the form of a linear trend test. In general, it will be assumed that the factor levels can be considered as equi-spaced.

For example, consider the variable ftv, which represents the number of visits to the gynecologist during the first trimester of pregnancy. This variable takes values between 0 and 6, the values greater than 2 being rarely observed (twelve cases in total, which(birthwt$ftv > 3)):

```
> table(birthwt$ftv)
```

```
   0    1    2    3    4    6
 100   47   30    7    4    1
```

One might choose to recode this counting variable into a qualitative variable with three levels, ftv2, after gathering the last three values (3, 4 and 6) with a value of 2 in a modality called 2+:

```
> birthwt$ftv2 <- birthwt$ftv
> birthwt$ftv2[birthwt$ftv2 > 2] <- 2
> birthwt$ftv2 <- factor(birthwt$ftv2, labels=c("0","1","2+"))
> table(birthwt$ftv2)
```

```
   0    1   2+
 100   47   42
```

This data can be represented in the form of a scatter diagram describing the variations of bwt according to the levels of ftv2 by using the command xyplot(). The option type=c("p","a") allows combining the representation of the individual observations in form of points ("p") with the averages for each level of the factor ftv2 ("a"). The option jitter.x=TRUE adds a slight horizontal offset to the points for increased readability in case of overlapping points (this does not affect the values read on the y-axis as the offset is horizontal):

```
> xyplot(bwt ~ ftv2, data=birthwt, type=c("p","a"),jitter.x=TRUE)
```

A conventional ANOVA model would give the following results:

```
> summary(aov(bwt ~ ftv2, data=birthwt))
            Df    Sum Sq  Mean Sq  F value  Pr(>F)
ftv2         2   1887925   943963     1.79    0.17
Residuals  186  98081730   527321
```

To test a linear trend effect between these two variables, two approaches can be used: the linear regression-based approach (the variable ftv2 is then regarded as a numeric variable and not a factor), which will be seen in more detail in the next chapter or that based on the utilization of contrasts associated to the levels of ftv2.

In the linear regression approach, we are considering that the explanatory variable is in fact a numeric variable and this is tantamount to performing a regression considering the variables bwt and ftv2 without converting the latter into a factor as has been done above. R can still be instructed to manipulate the qualitative variable ftv2 as a numeric variable using the command as.numeric(). The result of the linear regression, achieved by means of the command lm(), is summarized in the following ANOVA table. It indicates that the data are not compatible with the existence of a linear relationship between the variables bwt and ftv2:

```
> anova(lm(bwt ~ as.numeric(ftv2), birthwt))
Analysis of Variance Table

Response: bwt
                   Df    Sum Sq Mean Sq F value Pr(>F)
as.numeric(ftv2)    1    542578  542578  1.0205 0.3137
Residuals         187  99427078  531696
```

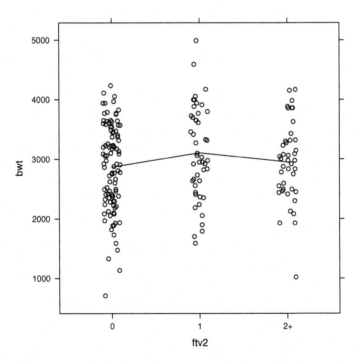

Figure 4.3. *Distribution of the weight of babies on the basis of the frequency of gynecological visits during the first quarter of pregnancy*

It should be noted that it can be easily verified how R has transformed the levels of ftv2 into numeric values:

```
> levels(birthwt$ftv2)
[1] "0"   "1"   "2+"
```

```
> sort(unique(as.numeric(birthwt$ftv2)))
[1] 1 2 3
```

Different levels of the factors are coded effectively as integers, with an offset of one unit between each level.

Regarding the method of contrasts, R will automatically provide this type of trend test when the qualitative variable is specified as a factor with ordered levels by using the command ordered() or as.ordered(). In case of a k-modality variable, R will automatically build $k - 1$ orthogonal contrasts making it possible to test the existence of a linear, quadratic, cubic, etc. relationship. In our example, the variable ftv2 has three levels and R will provide two contrasts: the first for the linear trend and the second for the quadratic trend (which amounts to consider the squared values of ftv2 in a regression model):

```
> birthwt$ftv3 <- ordered(birthwt$ftv2)
> res <- aov(bwt ~ ftv3, data=birthwt)
> summary(res, split=list(ftv3=c(Linear=1, Quadratic=2)))
                  Df   Sum Sq  Mean Sq  F value  Pr(>F)
ftv3               2  1887925   943963    1.790   0.170
  ftv3: Linear     1   542578   542578    1.029   0.312
  ftv3: Quadratic  1  1345348  1345348    2.551   0.112
Residuals        186 98081730   527321
```

As a note, a similar result can be obtained by using the command summary.lm() (this time, it is necessary to explicitly specify the suffix .lm) to obtain a summary of the coefficients of the linear model:

```
> summary.lm(res)
v
Call:
aov(formula = bwt ~ ftv3, data = birthwt)

Residuals:
     Min       1Q   Median       3Q      Max
-2156.14  -484.88    26.12   578.86  1882.00

Coefficients:
             Estimate Std. Error t value Pr(>|t|)
(Intercept)   2974.67      56.81  52.360   <2e-16 ***
ftv3.L          60.63      94.42   0.642    0.522
ftv3.Q        -163.29     102.23  -1.597    0.112
---
```

```
Signif. codes:  0 '***' 0.001 '**' 0.01 '*' 0.05 '.' 0.1 ' ' 1
```

```
Residual standard error: 726.2 on 186 degrees of freedom
Multiple R-squared:   0.01888,Adjusted R-squared:   0.008335
F-statistic:   1.79 on 2 and 186 DF,  p-value: 0.1698
```

4.3. Non-parametric one-way ANOVA

The non-parametric alternative to a single-factor ANOVA is the Kruskal-Wallis ANOVA, which consists of an ANOVA carried out on the ranks of the observations rather than their original values. The R command that can perform this type of analysis is kruskal.test(), and can be used in the same way as aov() using a formula in which the response variable and the explanatory variable are indicated. Unlike aov(), it is not necessary to combine the command kruskal.test() with summary().

Here is an example with birth weight data according to the ethnicity of the mother:

```
> kruskal.test(bwt ~ race, data=birthwt)

Kruskal-Wallis rank sum test

data:  bwt by race
Kruskal-Wallis chi-squared = 8.5199, df = 2, p-value = 0.01412
```

Pairwise treatment comparisons can be carried out by means of simple Wilcoxon tests, according to the same principle as presented in section 4.2.2.

4.4. Two-way ANOVA

In the presence of two factors, the formulation of the ANOVA model is substantially equivalent to that presented above. The relationship between variables is expressed by using R formula, the classification factors being located to the right of the symbol ~. This formula approach is derived from the symbolic notation for experimental design [WIL 73]. For example, y ~ x1 + x2 represents a model including two factors x1 and x2. The symbol : is used to bind the two variables to represent the interaction term. With the previous notations, y ~ x1 + x2 + x1:x2 would be used to represent the full ANOVA model (two main effects and one interaction effect). This last notation can be simplified in the form y ~ x1 * x2. In this context, the crossing of two modalities for each of the factors will be called "treatment".

By default, R makes use of Type-II sums of squares. In the case of balanced experimental design, the various methods for the calculation of the sums of squares provide equivalent results. To use other sums of squares, it is necessary to use the command drop1() or Anova() (package car).

4.4.1. Construction of an ANOVA table

When focusing on the effect of the ethnicity and of the history of hypertension in the mother on the weight of newborn babies, the ANOVA model can be formulated, including the interaction between these two factors, in the following manner:

```
> m <- aov(bwt ~ race + ht + race:ht, data=birthwt)
```

As it has been said, the statement aov(bwt ~ race * ht, data=birthwt) is strictly equivalent. The results of this model are shown here:

```
> summary(m)
            Df    Sum Sq Mean Sq F value  Pr(>F)
race         2   5015725 2507863   4.973 0.00789 **
ht           1   1776713 1776713   3.523 0.06211 .
race:ht      2    889649  444825   0.882 0.41568
Residuals  183  92287568  504304
---
Signif. codes:  0 '***' 0.001 '**' 0.01 '*' 0.05 '.' 0.1 ' ' 1
```

The ANOVA table indicates, for each variability source, the sum of squares associated, the degrees of freedom, the mean square and the F-test. We see that in this case, only the factor race is significant considering a 5% error risk. The interaction is not significant, which indicates that the effect of this factor does not depend on the level of ht.

The model without interaction is easily obtained by omitting the last term of the model, that is:

```
> summary(aov(bwt ~ race + ht, data=birthwt))
            Df    Sum Sq Mean Sq F value  Pr(>F)
race         2   5015725 2507863   4.979 0.00783 **
ht           1   1776713 1776713   3.528 0.06193 .
Residuals  185  93177217  503661
---
Signif. codes:  0 '***' 0.001 '**' 0.01 '*' 0.05 '.' 0.1 ' ' 1
```

In addition to a numerical summary for the treatment means, an interaction graph can better represent the results. It can be built using the following commands (Figure 4.4):

```
> xyplot(bwt ~ race, data=birthwt, groups=ht, type=c("a","g"),
      auto.key=list(corner=c(0,1), title="ht",
                    cex.title=1, points=FALSE, lines=TRUE))
```

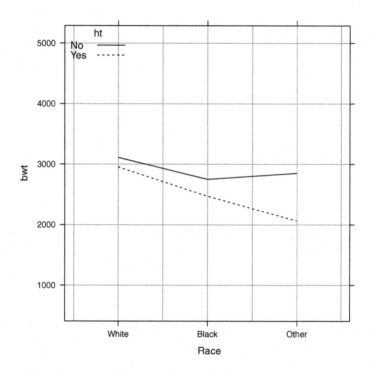

Figure 4.4. *Interaction diagram for the two-way ANOVA model*

For the sake of simplicity, we do not consider the addition of standard deviations or confidence intervals. The legend options being used make it possible to specify the variable utilized for the grouping and position of the legend (upper right corner), in which line segments are displayed (lines=TRUE) rather than points. The option type=c('a','g') enables the means of the treatments to be displayed (crossing the levels of the two factors) as well as grid lines which facilitate the identification of values being displayed.

To obtain the means of the treatments, one can either employ the command tapply(), or the command aggregate(). With the latter, the following would be written, using exactly the same formula:

```
> aggregate(bwt ~ race + ht, data=birthwt, mean)
   race  ht      bwt
1 White  No 3110.890
```

```
2 Black  No 2751.826
3 Other  No 2852.413
4 White Yes 2954.000
5 Black Yes 2473.333
6 Other Yes 2063.000
```

It should also be noted that R includes the command model.tables() which automatically calculates the average effects (deviations between the conditional means and the grand mean) as well as related standard errors:

```
> model.tables(m)
Tables of effects

 race
     White   Black   Other
     158.1  -224.9  -139.3
rep   96.0    26.0    67.0

 ht
          No    Yes
       25.15   -371
rep   177.00     12

 race:ht
          ht
race    No      Yes
  White  -12.6   229.6
  rep     91.0     5.0
  Black  -13.9   106.7
  rep     23.0     3.0
  Other   23.3  -367.0
  rep     63.0     4.0
```

4.4.2. Diagnostic model

Hypotheses of the two-way ANOVA model remain identical to those of the single-factor model (independence of the observations, normality and homogeneity of variances), except that they concern the residuals of the model and not the observed raw values. To test the hypothesis of homoscedasticity (equality of variances), for example, care will therefore be taken because Bartlett's test must include the six treatments formed from the interaction term, and not the levels of each factor separately. With the previous model, bwt ~ race * ht, the test should be formulated in the following manner:

```
> htrace <- with(birthwt, interaction(race, ht))
> bartlett.test(bwt ~ htrace, data=birthwt)

Bartlett test of homogeneity of variances

data:  bwt by htrace
Bartlett's K-squared = 5.1225, df = 5, p-value = 0.4011
```

The same decomposition principle according to the interaction term can be used to visualize the distribution of responses, of the predicted values or of the residuals depending on the treatments.

4.5. Key points

– rather than considering series of observations arranged in columns (the case of a balanced design), it is preferable to work with a table of data displaying a response variable explicitly and one or more classification factors;

– the command `aggregate()` can advantageously replace the command `tapply()` due to the use of R formulas;

– the command `aov()` allows an ANOVA model to be built, the table of the variance decomposition being provided directly by the command `summary()`.

4.6. Applications

1) In a study on the estrogen receptor gene, geneticists have focused on the relationship between the genotype and the diagnosis age of breast cancer. The genotype was determined from both alleles of a sequence restriction polymorphism (1.6 and 0.7 kb) or three groups of subjects: homozygous patients for the allele 0.7 kb (0.7/0.7), homozygous patients for the allele 1.6 kb (1.6/1.6) and heterozygous patients (1.6/0.7). The data were collected on 59 patients with breast cancer and are available in the file polymorphism.dta (Stata file) [DUP 09]. Average data are shown in Table 4.1.

	Genotype			
	1.6/1.6	1.6/0.7	0.7/0.7	Total
Number of patients	14	29	16	59
Age during the diagnosis				
Mean	64.64	64.38	50.38	60.64
Standard deviation	11.18	13.26	10.64	13.49
95% CI	(58.1-71.1)	(59.9-68.9)	(44.3-56.5)	

Table 4.1. *Estrogen receptor gene polymorphism*

a) test the null hypothesis according to which the age of diagnosis does not vary according to the genotype using an ANOVA. Represent in graphical form the distribution of ages for each genotype;

b) the confidence intervals presented in Table 4.1 have been estimated assuming the homogeneity of variances i.e. by using the estimate of the common variance; give the value of these confidence intervals without assuming the homoscedasticity;

c) estimate the differences on average corresponding to all the possible combinations of the three genotypes, with an estimation of the associated 95% confidence interval and a parametric test that allows the level of significance of the difference being observed to be evaluated;

d) represent graphically the group averages with 95% confidence intervals.

To load the data, it is necessary to import the library foreign that enables files saved by Stata to be read:

```
> library(foreign)
> polymsm <- read.dta("polymorphism.dta")
> head(polymsm)
```

Note that polymsm is a data frame, which will be helpful to use the graphics commands or to perform the ANOVA. The first column contains a series of unique identifiers for individuals; it will not be of any help in this case. To calculate the means and standard deviations for each group, one can proceed as follows:

```
> with(polymsm, tapply(age, genotype, mean))
> with(polymsm, tapply(age, genotype, sd))
```

The age distribution according to the genotype is shown in the following box plot:

```
> bwplot(age ~ genotype, data=polymsm)
```

It is also possible to employ histograms, by means of histogram() instead of bwplot():

```
> histogram(~ age | genotype, data=polymsm)
```

The ANOVA model is achieved through the command aov():

```
> aov.res <- aov(age ~ genotype, data=polymsm)
> summary(aov.res)
```

The F statistic is reported in the column F value, with the associated level of significance in the next column (Pr(>F)). The ANOVA indicates that at least a pair of means is significantly different, considering an error risk of 5%.

The accuracy of confidence intervals exposed in the table presented above can be verified. For this purpose, an estimate of the residual error is needed, which is simply the square root of the mean square associated to the error term in the previous ANOVA table:

```
> mse <- unlist(summary(aov.res))["Mean Sq2"]
> se <- sqrt(mse)
```

This yields an estimate of the root of the common variance of round(se,2). Therefore, 95% confidence intervals can be constructed for each group mean as follows:

```
> ni <- table(polymsm$genotype)
> n <- sum(ni)
> m <- with(polymsm, tapply(age, genotype, mean))
> lci <- m - qt(0.975, n-3) * se / sqrt(ni)
> uci <- m + qt(0.975, n-3) * se / sqrt(ni)
> rbind(lci, uci)
```

In general, for the estimation of the confidence interval of an average (or a proportion) one can refer to the command epi.conf() of the package epiR, for example:

```
> library(epiR)
> epi.conf(polymsm$age[polymsm$genotype=="1.6/1.6"],
           ctype = "mean.single")
```

To represent the group averages graphically, two approaches can be employed. The first consists of building the graph manually, presenting the averages in the form of points and the intervals of confidence in the form of segments:

```
> mm <- as.data.frame(cbind(m, lci, uci))
> mm$g <- levels(polymsm$genotype)
> rownames(mm) <- NULL
> dotplot(m ~ g, data=mm, ylim=c(40,75),
          panel=function(x, y, ...) {
              panel.dotplot(x, y, ...)
              panel.segments(x, mm$lci, x, mm$uci)
          })
```

The group averages were therefore grouped, as well as the upper and the lower bounds of the confidence intervals and the levels of the classification factor in the same data frame. The graphical command consists of appending a call to `panel.dotplot()` to display the averages, then `panel.segments()` to plot the associated confidence intervals.

The other easier solution consists of the command `xYplot()` (not to be confused with `xyplot()`) of the package `Hmisc` described in the appendix or `segplot()` of the package `latticeExtra`. In the first case, the only subtlety is that the explanatory variable (projected on the x-axis) should be treated as a numeric variable. The syntax has to be slightly adjusted to achieve the desired effect, in particular to properly display the labels corresponding to the different genotypes:

```
> xYplot(Cbind(m,lci,uci) ~ 1:3, data=mm,
         scales=list(x=list(at=1:3, labels=mm$g)),
         xlab="", ylim=c(40,75))
```

2) Any obstetrics service is concerned with the weight of newborn babies born at full term and infants aged one month [PEA 05]. For this sample of 550 babies, there is information available concerning parity (number of brothers and sisters), but it is known that there is no twinhood relationship among children having brothers and sisters. The purpose of the study is to determine if the parity (four classes) influences the weight of newborn babies in one month. The data are summarized in Table 4.2 and they are available in an SPSS file named `weights.sav`.

	Number of brothers and sisters				
	0	1	2	≥ 3	Total
Sample					
Counts	180	192	116	62	550
Proportion	32.7	34.9	21.1	11.3	100.0
Weight (kg)					
Mean	4.26	4.39	21.1	11.3	
Standard deviation	0.62	0.59	0.61	0.54	
(Min-Max)	(2.92-5.75)	(3.17-6.33)	(3.09-6.49)	(3.20-5.48)	

Table 4.2. *Weight of newborns*

a) Verify the data presented in the previous table.

b) Perform a one-way ANOVA. Draw conclusions about the global significance and report the part of variance explained by the model.

c) Display the weight distribution according to gender. Perform a homogeneity of variances test (search in the online help for the Levene test).

d) It was decided that the last two categories be combined (2 and $\geq$ 3). Carry out the analysis once more and compare to the results obtained in (b).

e) Perform a linear trend test (by ANOVA) with the data recoded into three levels for the parity.

To load the data, it is necessary to import the library foreign that enables reading files saved by SPSS:

```
> library(foreign)
> weights <- read.spss("weights.sav", to.data.frame=TRUE)
> str(weights)
```

As seen in previous exercises, the option to.data.frame= TRUE is important because it allows the data being read to be stored in form of a data frame (variables in columns, individuals in rows).

As the first step, we shall perform the numerical uni- and bivariate summary of the data of interest (variables WEIGHT and PARITY). Concerning the qualitative variable, the count and relative frequencies tables are obtained as follows:

```
> table(weights$PARITY)
> round(prop.table(table(weights$PARITY))*100, 1)
```

For the quantitative variable, the averages and the standard deviations by parity type are obtained as follows:

```
> round(with(weights, tapply(WEIGHT, PARITY, mean)), 2)
> round(with(weights, tapply(WEIGHT, PARITY, sd)), 2)
```

In the case of the ANOVA, the same principle as in the previous exercise is used:

```
> aov.res <- aov(WEIGHT ~ PARITY, data=weights)
> summary(aov.res)
```

The F-test is significant, and therefore the null hypothesis of the equality of the four averages can be rejected. The explained proportion of variance is simply the ratio between the sum of the squares (Sum Sq) associated with the study factor (PARITY), thus, 3.48 and the sum of the total squares or $3.48/(3.48 + 195.36)$: 0.018 is obtained approximately 2%.

To display the distribution of the weight it is possible to use box plots, as in the previous exercises. If the desired objective is that the individual data be directly visualized, a scatterplot conditioned on the groups is also interesting:

```
> stripplot(WEIGHT ~ PARITY, data=weights, ylab="Poids (kg)",
            jitter.data=TRUE)
```

The other alternative consists of using histograms for each group:

```
> histogram(~ WEIGHT | PARITY, data=weights)
```

It should be noted that a slight random offset has been added to the data (on the horizontal axis only) by making use of the option jitter.data=TRUE, which allows that the total overlapping of points be avoided.

Levene's test is not available in the basic R commands, but the package car can be installed, which provides the command leveneTest(), that can be used as any R testing procedure:

```
> library(car)
> leveneTest(WEIGHT ~ PARITY, data=weights)
```

An alternative would be to use a Bartlett test (bartlett.test()) for the homogeneity of variances. This command is used in exactly the same manner as leveneTest() (variable response described by the factor of study, name of the data frame where to search for the variables).

To group the last two categories, a new variable can be created and the new associated modalities manually generated, or more simply the qualitative variable PARITY can be recoded in a new variable:

```
> PARITY2 <- weights$PARITY
> levels(PARITY2)[3:4] <- "2 siblings or more"
```

The variance analysis model is constructed in the same way as before:

```
> aov.res2 <- aov(WEIGHT ~ PARITY2, data=weights)
> summary(aov.res2)
```

Finally, there are two solutions to perform a linear trend test: either by the method of contrasts or by a linear regression approach. These two approaches provide identical results and assume that the levels of the classification factor are equi-spaced (variation of the same number of units between each factor level). In this case, we will utilize the command lm() to perform a simple linear regression (see Chapter 5 for details):

```
> lm.res <- lm(WEIGHT ~ as.numeric(PARITY2), data=weights)
> summary(lm.res)
```

With respect to the ANOVA, the linear trend test is the test of the slope of the regression line, which is suggesting an increase in the average weight with the size of the siblings. Equivalently, one can use the approach consisting of recoding the classification factor into ordered modalities (or levels) factor using the command as.ordered(). The test for the linear trend corresponds to the contrast named as.ordered(PARITY2).L in the following output:

```
> summary(lm(WEIGHT ~ as.ordered(PARITY2), data=weights))
```

3) The data set ToothGrowth available in R contains data from a study on the length of odontoblasts (variable len) in ten guinea pigs after administration of vitamin C at different doses (0.5, 1 or 2 mg, variable dose) in the form of ascorbic acid or orange juice (variable supp) [BLI 52]:

a) verify the distribution of frequencies according to the different treatments (crossing the modalities of two factors, supp and dose) of this experimental design;

b) calculate the mean and the standard deviation of each treatment;

c) build an ANOVA table for the full model including the interaction between the two factors;

d) draw an interaction chart that represents the average values of the response variable according to the levels of the two explanatory variables;

e) verify that the homogeneity of variances assumption is acceptable for these data.

To load the data, the command `data()` will be used since data are provided with R:

```
> data(ToothGrowth)
> str(ToothGrowth)
```

The examination of the coding of the variables using `str()` indicates that the variables supp and dose are currently treated as numeric variables. In the case of the variable supp, this does not raise any problems because it has just two distinct levels. On the other hand, the variable dose must be recoded into a factor if the ANOVA models are to be used:

```
> ToothGrowth$dose <- factor(ToothGrowth$dose)
```

Here, the modalities of dose are considered unordered.

To verify the frequency distribution according to the different modalities of this factorial design, it is quite possible to use `table()` or `xtabs()`, for example:

```
> xtabs(~ dose + supp, data=ToothGrowth)
```

However, R also provides a command that makes it possible to summarize the frequencies of each factor and the interaction between these two factors (which corresponds to the numbers associated with each treatment when the experimental design is balanced). In the first instance, the type of relationship that interests us can be indicated using an R formula. Here, this corresponds to the ANOVA model which attempts to explain the variations in the response variable, len, according to the levels of the variables supp and dose, these two factors being considered in interaction:

```
> fm <- len ~ supp * dose
> replications(fm, data=ToothGrowth)
```

The relevance of using a formula is that it can be reused to calculate the averages by treatment and the ANOVA model. The symbol * in the formula is used to expresses that the two factors are interacting in the model.

In the following, it is shown how the means and standard deviations can be calculated, after defining a small R function that can calculate these two statistics from a series of observations:

```
> f <- function(x) c(mean = mean(x), sd = sd(x))
> aggregate(fm, ToothGrowth, f)
```

The ANOVA model raises no particular difficulty and the ANOVA table will be obtained by combining `aov()` and `summary()`:

```
> aov.fit <- aov(fm, data = ToothGrowth)
> summary(aov.fit)
```

By means of the command `model.tables()`, it is possible to construct a table of the simple effects in this model. In this case, the values returned by R correspond to the differences between the averages for each factor level and the "grand mean" or overall mean:

```
> model.tables(aov.fit, type = "means", se = TRUE,
               cterms = "supp:dose")
```

Regarding the interaction graph, a slightly adapted simple scatter plot will be used:

```
> xyplot(len ~ dose, data = ToothGrowth, groups = supp,
         type = c("a", "g"), xlab = "Dose (mg)",
         ylab = "Tooth length", lwd = 2,
         auto.key = list(space = "top", columns = 2,
                         points = FALSE, lines = TRUE))
```

The option `type="a"` allows the automatic calculation of the averages conditionally to the levels of the factors included in the composition of the graph. R also provides a command `interaction.plot()` among the basic graphics.

The distribution of the observations for each treatment makes it possible to visually inspect if the assumption of homogeneity of variances seems acceptable or not. However, the treatments are effectively what matter here (since the ANOVA model includes the terms of interaction between the two factors). Consequently, the fact that the distribution of the response variable is displayed in the form of a boxplot assumes that we are effectively working with the treatments. To this end, it is possible to have recourse to the command `interaction()` that enables that all the levels of each factor be combined:

```
> bwplot(len ~ interaction(supp, dose), data = ToothGrowth,
         do.out = FALSE, pch = "|")
```

A formal test of equality of the variances is available through the command `bartlett.test()`:

```
> bartlett.test(len ~ interaction(supp, dose), data=ToothGrowth)
```

5

Correlation and Linear Regression

In this chapter, we will focus on monotonic or linear association measures between two numeric variables (pointwise or interval estimation, correlation coefficient test) and on the associated exploratory graphical representations (scatter diagram and Loess curve). The linear regression model is then developed in the case of an explanatory variable (simple linear regression); particularly the estimation of the regression line coefficients and the construction of the associated null hypothesis tests, the table of the variance analysis associated with the regression model, as well as pointwise and interval prediction. The verification of the application conditions and the model diagnosis by analysis of the residuals are also considered.

5.1. Descriptive statistics

The descriptive summaries for two numeric variables are easily obtained with summary() or a command operating by variable (sapply(), seen in Chapter 4, or apply()), in the case where the variables are arranged in columns. When focusing on the weight measures (lwt) of the mothers and of the newborns (bwt), the following commands essentially provide the same information:

```
> summary(birthwt[,c("bwt","lwt")])
> summary(birthwt[,c(10,3)])
> summary(subset(birthwt, select=c(bwt, lwt)))
> sapply(birthwt[,c("bwt","lwt")], summary)
> apply(birthwt[,c(10,3)], 2, summary)
         bwt    lwt
Min.     709   80.0
```

```
1st Qu. 2414 110.0
Median  2977 121.0
Mean    2945 129.8
3rd Qu. 3487 140.0
Max.    4990 250.0
```

A few observations: it is always easier to read R scripts when the variables are identified explicitly (c("bwt","lwt")) rather than addressed by their position in the table (c(10,3)); the command apply() has the advantage of working also with matrices (rectangular arrays composed of variables of the same type); the name of the variables in the command subset() must not be surrounded by English quotes.

It is possible to visualize the histograms of these two variables of interest by using the command histogram() directly (Figure 5.1). The only subtlety is that in general the commands of the package lattice aim to unify the units of measurement for the abcissa and ordinate axes. In this case, this means that histogram() will build an x-axis representing the largest variations (babies' weight expressed in grams, within a range of 700 to 5,000) which will make the histogram of the mothers' weight unreadable. To prevent this, it is necessary to force R to build two different coordinate systems, through the option scales=, in the following manner:

```
> histogram(~ bwt + lwt, data=birthwt, breaks=NULL, xlab="",
            scales=list(x=list(relation="free")))
```

In the absence of a classification variable (factor) allowing two sets of measures to be separated, it is sufficient to indicate the list of variables whose distribution is to be displayed in the form of a histogram separated by a +.

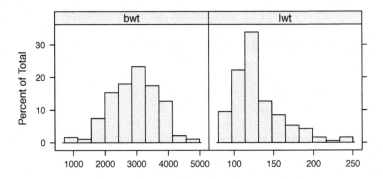

Figure 5.1. *Histograms of the mothers' (lwt, in kg) and the newborns' weight (bwt, in g), presented as percentages*

5.1.1. *Scatterplot and Loess curve*

Before calculating any statistic on a bivariate series of observations, it is useful to visualize the relationship between the two variables graphically, most often in the form of a scatterplot. The command xyplot() has a number of graphical options with the option type= which facilitates the identification of a more or less linear association between two variables, in particular the option type="smooth". This allows overlaying a curve of the "Loess" type [CLE 79] on the scatter plot that is based on local regression, whose degree of smoothing can be controlled (see help(loess)).

Firstly, we will convert the mothers' weight in kilograms, then we will display the scatterplot and the Loess curve:

```
> birthwt$lwt <- birthwt$lwt/2.2
> xyplot(bwt ~ lwt, data=birthwt, type=c("p","g","smooth"))
```

The degree of smoothing of the Loess curve can be controlled with the option span=; for example, span=1 will provide a curve which is less sensitive to local variations and thus with a much smoother appearance (Figure 5.2).

5.1.2. *Parametric and non-parametric association measures*

The Pearson correlation coefficient is obtained from the command cor(), indicating the two variables of interest:

```
> with(birthwt, cor(bwt, lwt))
[1] 0.1857333
```

In the case where data is missing, it is necessary to indicate to R how to calculate the correlation coefficient. A conventional approach (in the bivariate case or to build correlation matrices on a larger number of variables) consists of considering only the pairs of complete observations. If there are two variables x and y, R will exclude the statistical units in which is.na(x) or is.na(y) is equal to TRUE (this obviously includes the conjunction of the two conditions). The option to be included in the command cor is use="pairwise".

To calculate the Spearman correlation coefficient (based on the ranks of the observations), the option method="spearman" will be added:

```
> with(birthwt, cor(bwt, lwt, method="spearman"))
[1] 0.2488882
```

These two commands also operate with more than two variables: in this case, it is sufficient to provide to cor() a table with the variables of interest, for example:

```
> cor(birthwt[,c("age","lwt","bwt")])
          age       lwt       bwt
age 1.00000000 0.1800732 0.09031781
lwt 0.18007315 1.0000000 0.18573328
bwt 0.09031781 0.1857333 1.00000000
```

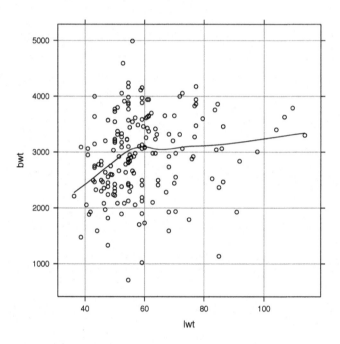

Figure 5.2. *Scatter plot representing the variations of the weight of newborns (bwt) according to the weight of the mothers (lwt)*

5.1.3. *Interval estimation and inference test*

The command cor.test() provides both a test of the hypothesis $H_0: \rho = 0$ (parameter equal to zero, that is theoretical correlation equal to zero) through a Student's t-test, as well as a confidence interval calculated by Fisher inverse transformation. In the nonparametric case, R proposes to calculate an exact degree of significance for samples of size $n < 50$ (exclusively); otherwise, a normal distribution approximation is being employed:

```
> with(birthwt, cor.test(bwt, lwt))
```

```
Pearson's product-moment correlation

data:  bwt and lwt
t = 2.5848, df = 187, p-value = 0.0105
alternative hypothesis: true correlation is not equal to 0
95 percent confidence interval:
 0.04417405 0.31998094
sample estimates:
      cor
0.1857333
```

It should be noted that a formula notation can be included with this command, in a fashion similar to that used for xtabs() when both variables have a symmetrical function. It is also possible to restrict the observations used for the estimation of the parameter and when testing if it is equal to zero by using the option subset=. For example:

```
> cor.test(~ bwt + lwt, data=birthwt, subset=bwt > 2500)

Pearson's product-moment correlation

data:  bwt and lwt
t = 1.6616, df = 128, p-value = 0.09905
alternative hypothesis: true correlation is not equal to 0
95 percent confidence interval:
 -0.02757158  0.30974090
sample estimates:
      cor
0.1453043
```

will be used to restrict the calculation to newborn babies with weights greater than 2.5 kg only.

The option method="spearman" will be added to perform an equivalent inference test on the correlation coefficient based on the ranks of the observations.

5.2. Simple linear regression

5.2.1. Regression line

To achieve a linear regression and more generally any linear model (multiple linear regression, analysis of variance, analysis of covariance), R proposes the command lm(). Its usage is essentially identical to that of aov() seen in Chapter 4:

using an R formula, the relationship between variables is described (response variable and explanatory variable), the data frame in which the variables are available, and possibly the subset of observations intended to be employed with the option subset=.

To model the relationship between the weight of the newborns (response variable) and the weight of the mothers (explanatory variable or predictor), the following formulation shall be used:

```
> lm(bwt ~ lwt, data=birthwt)

Call:
lm(formula = bwt ~ lwt, data = birthwt)

Coefficients:
(Intercept)           lwt
  2369.624         9.744
```

R returns the y-intercept ("Intercept"), the value of bwt when lwt=0, and the slope of the regression line; the latter reflecting the variations in bwt when lwt varies by one unit.

As seen in the case of the ANOVA with aov() (section 4.2.1), the information returned by the command lm is minimal. To obtain the regression coefficient table, their standard error and their degree of significance resulting from Student's t-tests, the command lm() will again be associated to summary():

```
> r <- lm(bwt ~ lwt, data=birthwt)
> summary(r)

Call:
lm(formula = bwt ~ lwt, data = birthwt)

Residuals:
    Min       1Q    Median       3Q       Max
-2192.12  -497.97     -3.84   508.32   2075.60

Coefficients:
            Estimate Std. Error t value Pr(>|t|)
(Intercept) 2369.624    228.493  10.371   <2e-16 ***
lwt            9.744      3.770   2.585   0.0105 *
---
Signif. codes:  0 '***' 0.001 '**' 0.01 '*' 0.05 '.' 0.1 ' ' 1
```

```
Residual standard error: 718.4 on 187 degrees of freedom
Multiple R-squared:  0.0345,Adjusted R-squared:  0.02933
F-statistic: 6.681 on 1 and 187 DF,  p-value: 0.0105
```

In addition to the significance tests on the regression coefficients, R includes additional information, such as the residual variance and the coefficient of determination R^2, that reflects the proportion of variance explained by the regression (see section 5.2.2).

It is possible to include the regression line in the scatter plot, as with the Loess curve, including the option type="r" (instead of type="smooth"), although the two options can be combined together (Figure 5.3):

```
> xyplot(bwt ~ lwt, data=birthwt, type=c("p","g","r"))
```

The default symbol (circle) is appropriate in the case of large numbers of observations, but the size can either be adjusted (cex=0.8, for example), or transparent symbols used (pch=19, alpha=0.5).

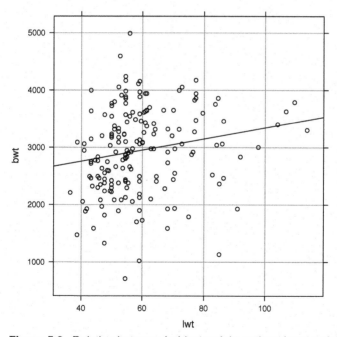

Figure 5.3. *Relation between babies' weight and mothers' weight and associated regression line*

5.2.2. *Interval estimation and variance analysis table*

The command `summary()` applied to a regression model does not provide confidence intervals for the regression coefficients. To this end, use the command `confint()`, optionally specifying the width of the confidence interval with the option `level=` (by default, it is equal to 0.95 for an 95% interval):

```
> confint(r)
                2.5 %      97.5 %
(Intercept) 1918.867879 2820.37916
lwt            2.307459   17.18061
```

These intervals could be associated to the coefficients estimated in the previous stage by constructing a three-column table:

```
> res <- cbind(coef(r), confint(r))
> colnames(res)[1] <- "Coef"
> round(res, 3)
                Coef     2.5 %     97.5 %
(Intercept) 2369.624 1918.868 2820.379
lwt            9.744    2.307   17.181
```

Here, we use the fact that the command `coef()` returns the regression coefficients estimated in the model named `r`, while `cbind()` allows lists of numbers to be concatenated in columns (provided that each list contains the same number of elements).

The ANOVA table associated with the regression model allows the decomposition of the total variance in two parts: the variance explained by the model and the residuals. It is obtained by using the command `anova()`:

```
> anova(r)
Analysis of Variance Table

Response: bwt
           Df   Sum Sq  Mean Sq F value  Pr(>F)
lwt         1  3448639  3448639  6.6814  0.0105 *
Residuals 187 96521017   516155
---
Signif. codes:  0 '***' 0.001 '**' 0.01 '*' 0.05 '.' 0.1 ' ' 1
```

Essentially, the same results as those returned by the command `summary()` applied to the output of the command `lm()` are obtained.

5.2.3. *Regression model predictions*

As in the case of the command `aov()` for the ANOVA, the two commands `fitted()` and `predict()` are used to obtain the adjusted values and build new predictions respectively. The command `fitted(r)` will return the predicted values for the weight of the observed babies. In this case, it is generally referred to as "fitted values":

```
> head(fitted(r))
      85       86       87       88       89       91
3175.721 3056.135 2834.680 2847.967 2843.538 2918.833
```

It is easy to compare the observed values with those predicted by the model (located on the right of the regression line) by juxtaposing the two sets of values in the following manner:

```
> head(cbind(birthwt$bwt, fitted(r)))
     [,1]     [,2]
85   2523 3175.721
86   2551 3056.135
87   2557 2834.680
88   2594 2847.967
89   2600 2843.538
91   2622 2918.833
```

It is also possible to add the predicted values to the previous scatter plot by slightly modifying the R formula being used (Figure 5.4): in this case, two variables will be shown on the left of the operator ~ to instruct R that there are two sets of measures to represent with respect to the same variable projected on the x-axis (here, `lwt`):

```
> xyplot(bwt + fitted(r) ~ lwt, data=birthwt)
```

The command `predict()` is used to obtain the values predicted from one set of observations other than those observed in the working data. It is then necessary to indicate to R the values taken by the explanatory variable, here `lwt`, and to store them in a data frame by using the option `newdata=`. Here follows an example where the values predicted by the regression model are estimated for maternal weights ranging from 35 to 120 kg, in 5 kg increments:

```
> dp <- data.frame(lwt=seq(35, 120, by=5))
> bwtpred <- predict(r, newdata=dp)
> dp$bwt <- bwtpred
> head(dp)
  lwt      bwt
```

```
1   35  2710.665
2   40  2759.385
3   45  2808.105
4   50  2856.825
5   55  2905.546
6   60  2954.266
```

It is possible to add the option se=TRUE to obtain the standard error associated with each predicted value and interval="prediction" for a 95% confidence interval for the prediction. With regard to the values involved in the construction of the regression model, interval="confidence" might be sufficient to build 95% confidence intervals for the values predicted (fitted) from the actual data.

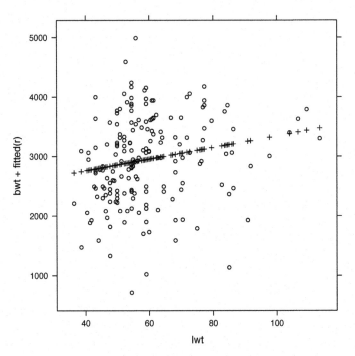

Figure 5.4. *Data observed (o) and predicted (+) by the linear regression model*

5.2.4. *Diagnostic and residual analysis of the model*

The coefficient of determination is a good indicator of the quality of the model, but it provides no information on the possible observations that may have influenced its parameters (regression coefficients). To this end, R offers a whole range of influence measures for the observations used in the model through the command influence.measures(). This includes the DFBETAS values for the two regression coefficients, the DFFIT values or the Cook distances [FOX 10, DRA 98]. It is also possible to calculate the standardized or studentized residuals of a regression by making use of the commands rstandard() (or stdres() in the package MASS) and rstudent() (or studres() in the package MASS).

Similarly to fitted(), the command resid() enables the value of the residuals (deviations between the observed values and the predictions) of the regression model to be obtained:

```
> head(resid(r))
       85        86        87        88        89        91
-652.7211 -505.1352 -277.6798 -253.9671 -243.5380 -296.8329
> head(birthwt$bwt - fitted(r))
       85        86        87        88        89        91
-652.7211 -505.1352 -277.6798 -253.9671 -243.5380 -296.8329
```

A graphical representation that is useful for verifying the stability of the variance is to display the values of the residuals according to the values observed or predicted by the model (Figure 5.5). For example:

```
> xyplot(resid(r) ~ fitted(r), type=c("p","g"))
```

To trace a horizontal line crossing the vertical axis at 0, it is possible to add the option abline=list(h=0, lty=2), for example. The package lattice provides the command rfs() which essentially generates the same result.

The other assumptions of the linear regression model (linearity and normality of errors) can be verified using a simple scatterplot (with a Loess curve, for example) and a histogram of the residuals.

5.2.5. Connection with ANOVA

The command aov() is based mainly on lm() but in addition, it allows that specific error terms be built for the statistical tests; in the cases where the observations are not independent (case of the so-called "repeated measures" protocols, for example). On the other hand, it is possible to study the relationship between a numerical response variable and a categorical predictor with a regression approach: the results will be approximately interpreted in the same way. Without going into the details of the coding of contrasts and their manipulation in R, the results of a linear regression can be seen, considering the variables bwt and race, which have been studied in Chapter 4 (section 4.2.1):

```
> lm(bwt ~ race, data=birthwt)

Call:
lm(formula = bwt ~ race, data = birthwt)

Coefficients:
(Intercept)     raceBlack     raceOther
     3102.7        -383.0        -297.4
```

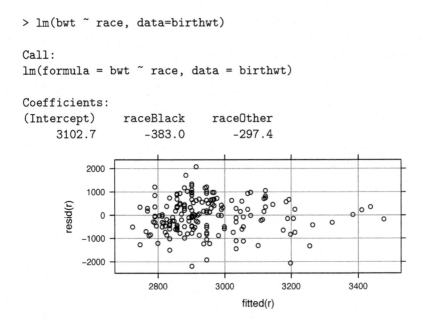

Figure 5.5. *Graph of the fluctuation of the residuals
according to the fitted values*

The preceding command returns three coefficients: a term for the y-intercept and two regression coefficients corresponding to the last two levels of the factor race. Since R relies on "treatment contrasts" by default, these results are to be interpreted in the following manner: the y-intercept represents the average weight of babies for mothers whose ethnicity is White, while the other two terms represent the differences between the average weights for the groups Black and Other and the average weight of babies in the group White. It is quite simple to numerically verify it:

```
> bwtmeans <- with(birthwt, tapply(bwt, race, mean))
> bwtmeans
    White     Black     Other
3102.719 2719.692 2805.284
> bwtmeans[2:3] - bwtmeans[1]
    Black     Other
-383.0264 -297.4352
```

It should be noted that the y-intercept term could very well be suppressed to find the three group averages. To this end, it is necessary to modify the R formula slightly:

```
> lm(bwt ~ race - 1, data=birthwt)
```

```
Call:
lm(formula = bwt ~ race - 1, data = birthwt)

Coefficients:
raceWhite  raceBlack  raceOther
     3103       2720       2805
```

The notation bwt ~ race + 0 is also valid.

Regardless of the interpretation of the regression coefficients, it is also possible to verify whether the ANOVA table for the regression provides much of the same information as that of the one-way ANOVA:

```
> anova(lm(bwt ~ race, data=birthwt))
Analysis of Variance Table

Response: bwt
          Df   Sum Sq Mean Sq F value   Pr(>F)
race       2  5015725 2507863  4.9125 0.008336 **
Residuals 186 94953931  510505
---
Signif. codes:  0 '***' 0.001 '**' 0.01 '*' 0.05 '.' 0.1 ' ' 1
```

5.3. Multiple linear regression

Regarding multiple regression, the usage of lm() remains the same: the response variable is placed to appear on the left side of the operator ~ and on its right side the explanatory variables that one wants to include in the model. In the case of an additive model, we will for example use a formula like lm(y ~ x1 + x2) to indicate to R to calculate the (partial) regression coefficients associated with x1 and x2. The notation x1:x2 to designate an interaction remains valid, and in the case where one of the two variables is categorical this corresponds to the covariance analysis model (of which a detailed example can be found in [VEN 02]).

It is sometimes helpful to center a numeric variable, particularly when it is involved in an interaction term with a qualitative variable; this can be achieved by using the command scale(). For example, the command:

```
> lm(bwt ~ scale(lwt, scale=FALSE) + ftv, data=birthwt)
```

allows the estimation of the effect of the mother's weight (lwt, centered on its mean) and of the frequency of visits to the gynecologist in the first trimester of pregnancy. Since the normalization of the observations by the standard deviation is not desired, it is necessary to specify the option scale=FALSE when using scale().

Finally, the variables (response or explanatory) can be perfectly transformed when the R formula linking the variables is built. For example, `y ~ log10(x1) + x2` would model the logarithm (base 10) of `x1` instead of `x1` directly. It should be noted that the inclusion of quadratic terms, cubic or other, should be achieved as follows: `y ~ x 1 + I(x1^2) + x2`, to indicate to R that the transformation `^2` applies to the values of `x1` and should not be interpreted as a formula term (in this case, the notation `^2` enables the automatic building of second-order interaction terms).

5.4. Key points

– Before estimating a correlation or a regression coefficient assuming a linear relationship between two variables, the joint distribution of two variables will be graphically represented using a histogram (`histogram()` with an option `type="smooth"`).

– The commands `cor()` and `cor.test()` are used to calculate the Pearson and Spearman correlation coefficients (`method="spearman"`), and to test the parameter for the null hypothesis.

– The command `lm()` makes it possible to build a linear regression model; it is employed with `summary()` to obtain the table of the regression coefficients and with `anova()` for the associated ANOVA table.

5.5. Going further

The packages `car` [FOX 10] and `rms` [HAR 01] provide a set of commands specific to linear regression which allow, among other things, to verify whether the conditions for the application of the single or multiple linear regression model are satisfied. The works of Julian Faraway and John Fox [FAR 14, FOX 10] provide an in-depth discussion and numerous illustrations of data modeling by means of linear (and nonlinear) models.

5.6. Applications

1) A study has focused on a malnutrition measurement among 25 patients aged from 7 to 23 years old and suffering from cystic fibrosis. Different information about the anthropometric characteristics (height, weight, etc.) and the pulmonary function of these patients [EVE 01] was available. The data are available in the file `cystic.dat`:

a) calculate the linear correlation coefficient between the variables `PEmax` and `Weight`, as well as its 95% confidence interval;

b) display the numerical data in the form of scatterplots i.e. 45 graphs arranged in the form of a "dispersion matrix";

c) calculate the Pearson and Spearman correlations between the numerical variables;

d) calculate the correlation between PEmax and Weight, controlling the age (Age) (partial correlation). Graphically represent the covariation between PEmax and Weight highlighting the two most extreme terciles for the variable Age.

The data being available in a text format in which the values ("fields") are separated by tabs, the command read.table() will be used to import them into R. Since the first line in the file enables the identification of the variables, the option header=TRUE will be added:

```
> cystic <- read.table("cystic.dat", header=TRUE)
> str(cystic)
> summary(cystic)
```

It can be easily seen that the gender of the patients is not encoded in the form of a qualitative variable but of a number (0/1). Although this is not fundamentally necessary in this exercise, it is always preferable to convert the variables into the correct format:

```
> cystic$Sex <- factor(cystic$Sex, labels=c("M","F"))
> table(cystic$Sex)
```

The linear correlation between PEmax and Weight is obtained by using the command cor(), which by default calculates a Pearson correlation coefficient:

```
> with(cystic, cor(PEmax, Weight))
```

The somewhat strange construction with(cystic,...) makes it possible to avoid repeating the name of the data frame, cystic, to denote the variables of interest. Otherwise, it would have been written: cor(cystic$PEmax, cystic$Weight). The command cor() only allows the pointwise estimation of the parameter. To obtain the associated confidence interval, the command cor.test() will be used directly, which at the same time provides an answer to the null hypothesis test of the parameter:

```
> with(cystic, cor.test(PEmax, Weight))
```

To display the scatterplots, the basic command `pairs()` or the following solution can be used:

```
> splom(cystic[,-c(1,3)], varname.cex=0.7, cex=.8)
```

Note that we have excluded the identifier of the subjects as well as the variable Sex that appears in the third column of the data table. The other two options (`varname.cex` and `cex`) allow the controlling of the size of the fonts and points.

To calculate the correlations of all pairs of variables, the command `cor()` can always be used. It will not be possible to test these correlations with the command `cor.test()`; to this end the command `corr.test()` of the package psych would be inserted:

```
> round(cor(cystic[,-3]), 3)
```

Concerning the Spearman correlations, the option `method="spearman"` is added as follows:

```
> round(cor(cystic[,-3], method="spearman"), 3)
```

Finally, to estimate the partial correlation between PEmax, Weight and Age, the command `pcor.test()` of the package ppcor will be applied:

```
> library(ppcor)
> with(cystic, pcor.test(PEmax, Weight, Age))
```

The controlling variable is indicated in the third position. It is possible to verify that the correlation between PEmax and Weight strongly decreases when the correlation between these two variables and age is considered using the following:

```
> cystic$Age.ter <- cut(cystic$Age,
                    breaks=quantile(cystic$Age,
                                c(0,0.33,0.66,1)),
                    include.lowest=TRUE)
> cystic2 <- subset(cystic, as.numeric(Age.ter) %in% c(1,3))
> cystic2$Age.ter <- factor(cystic2$Age.ter)
> xyplot(PEmax ~ Weight, data=cystic2, groups=Age.ter,
        auto.key=list(corner=c(0,1)))
```

2) In the Framingham study, data about systolic blood pressure (sbp) and about the index of body mass (bmi) are available for 2,047 men and 2,643 women [DUP 09]. The main focus is the relationship between these two variables (after logarithmic transformation) in men and in women separately. The data are available in the file Framingham.csv:

a) graphically represent the variations between blood pressure and BMI (bmi) in men and in women;

b) are the linear correlation coefficients estimated for men and for women significantly different at 5%?

c) estimate the parameters of the linear regression model considering the blood pressure as a response variable, and the BMI as an explanatory variable for these two subsamples. Give a 95% confidence interval for the estimation of the respective slopes.

This time, data was generated from a spreadsheet (Excel or similar) but the field delimiter is the "," hence the use of read.csv() instead of read.csv2() since the default options correspond to the case to be addressed:

```
> fram <- read.csv("data/Framingham.csv")
> head(fram)
> str(fram)
```

The variable sex is considered a quantitative variable (1/2) and to facilitate the interpretation, we immediately recode it into a qualitative variable:

```
> table(fram$sex)
> fram$sex <- factor(fram$sex, labels=c("M","F"))
```

Before considering the variations between blood pressure and BMI, it may be desirable to check for missing values. This does not affect the estimation of the means and standard deviations of the parameters of the regression line, or the graphical representation of the data, but it enables us to know the number of "complete cases" with the data being considered:

```
> apply(fram, 2, function(x) sum(is.na(x)))
```

A command apply() is used to repeat the same operation for each variable; this operation consists of counting (sum) the number of missing values (is.na(x)). We observe that for one of the variables in our model, the BMI (bmi), nine observations are missing. Hence the following distribution by gender:

```
> with(fram, table(sex[!is.na(bmi)]))
```

To represent the data in a scatterplot, due to the "large values" by subgroup (more than 2,000 observations), there is a risk of many points overlapping. One possibility is to use semi-transparency and to reduce the size of the points:

```
> xyplot(sbp ~ bmi | sex, data=fram, type=c("p","g"), alpha=0.5,
        cex=0.7, pch=19)
```

The correlations between the variables sbp and bmi in men and women can be obtained with the command cor(), naturally restricting the analysis to each of the two subsamples:

```
> with(subset(fram, sex=="M"), cor(sbp, bmi, use="pair"))
> with(subset(fram, sex=="F"), cor(sbp, bmi, use="pair"))
```

It should be noted that the option use="pair" has been added (the complete option is read as "pairwise.complete.obs" but it is possible to abbreviate the options when this does not raise any problems of ambiguity) to calculate the correlations of the set of observed data. To test if these two coefficients estimated from the data can be regarded as significantly different at the 5% threshold, the command r.test() of the package psych has to be used:

```
> library(psych)
> r.test(n=2047, r12=0.23644, n2=2643, r34=0.37362)
```

To verify the effect of the logarithmic transformation on the distribution of the variables sbp et bmi, two approaches can be followed: either create a data frame containing the density values of each histogram, associated to the type of variable (bmi, log(bmi), sbp and log(sbp)), or separately build four histograms and combine them into a single figure by making use of the command grid.arrange() of the package gridExtra. The following is obtained with this second solution:

```
> library(gridExtra)
> p1 <- histogram(~ bmi, data=fram)
> p2 <- histogram(~ log(bmi), data=fram)
```

```
> p3 <- histogram(~ sbp, data=fram)
> p4 <- histogram(~ log(sbp), data=fram)
> grid.arrange(p1, p2, p3, p4)
```

The regression model in each subgroup and the 95% confidence interval associated with the regression slopes can be calculated as:

```
> reg.resM <- lm(log(sbp) ~ log(bmi), data=fram, subset=sex=="M")
> reg.resF <- lm(log(sbp) ~ log(bmi), data=fram, subset=sex=="F")
> summary(reg.resM)    # Hommes
> confint(reg.resM)
> summary(reg.resF)    # Femmes
> confint(reg.resF)
```

Regarding the confidence intervals, the command confint() is used. All the results that we are interested in can be grouped in a same table. For example:

```
> res <- data.frame(pente=c(coef(reg.resM)[2],
                            coef(reg.resF)[2]),
                    rbind(confint(reg.resM)[2,],
                          confint(reg.resF)[2,]))
> rownames(res) <- c("M","F")
> colnames(res)[2:3] <- c("2.5 %", "97.5 %")
> round(res, 3)
```

3) Based on the data on weight at birth [HOS 89], the aim is to study the relationship between the weight of babies (considered as a numeric variable, bwt) and two characteristics of the mother: her weight (lwt) and her ethnic origin (race):

a) graphically represent the relationship between babies' weight and mothers' weight, according to the ethnicity of the mothers;

b) estimate the linear regression parameters, considering the baby's weight as a response variable and the mothers' weight centered on their average as an explanatory variable. Is the estimated slope significant at the usual threshold of 5%?

c) estimate the parameters of the linear regression, this time considering the mothers' ethnicity as the explanatory variable (the baby's weight remains the response variable). Compare the significance of the model with the results obtained from a one-way ANOVA (ethnicity);

d) what is the weight predicted for a baby whose mother weighs 60 kg?

e) give a 95% confidence interval for a pointwise prediction on average;

f) perform once more the regression analysis described in (c) after modifying the way R generates the contrasts for the qualitative variables by typing the following at the R command prompt:

```
> options(contrasts=c("contr.sum", "contr.poly"))
```

Compare with the previous results.

The data can be imported and recoded exactly as in Chapter 1 (section 1.5.2).

The relation between babies' weight and mothers' weight can be represented by using a simple scatterplot. Here, it has been decided to stratify the ethnicity of mothers and to display three separate charts:

```
> xyplot(bwt ~ lwt | race, data=birthwt, layout=c(3,1),
        type=c("p","g"), aspect=0.8)
```

The option `aspect=0.8` allows the width-height ratio of the graphs to be modified. The option `layout=c(3,1)` enables the display of the three graphs side by side (one row, three columns). By slightly modifying the previous command, it would be possible to display all the observations in the same chart but using different symbols or colors depending on the ethnic origin:

```
> xyplot(bwt ~ lwt, data=birthwt, groups=race)
```

The parameters of the linear regression model are estimated with the command `lm()`, knowing that the command `scale()` makes it possible to center and/or reduce a series of measures. Here, we only want to center the mothers' weights on their average, and not to standardize them by units of standard deviation:

```
> reg.res <- lm(bwt ~ scale(lwt, scale=FALSE), data=birthwt)
> summary(reg.res)
```

The t-test assessing the null hypothesis for the slope is significant ($p = 0,011$) if a 5% Type-I error risk is being considered.

Considering the ethnicity as an explanatory variable, the regression model is estimated in the same manner:

```
> reg.res2 <- lm(bwt ~ race, data=birthwt)
> summary(reg.res2)
```

It is assumed that the explanatory variable is to be actually re-encoded in the form of a factor as in Chapter 1. On the other hand, the ANOVA would give us:

```
> aov.res <- aov(bwt ~ race, data=birthwt)
> summary(aov.res)
```

In fact, it is possible to obtain the previous regression table by using summary.lm() instead of summary(), which can be verified using the following output:

```
> summary.lm(aov.res)
```

Conversely, an ANOVA table corresponding to the previous regression model can be displayed:

```
> anova(reg.res2)
```

It can be verified that the F-statistic is the same in both cases (it is equal to 4.913, 2 to 186 degrees of freedom).

The contrasts used in the regression model (reg.res2) are called treatment contrasts, and they allow us to test the difference between the mean scores (here, the age) of two categories of a qualitative variable (here, the ethnicity), one of the two categories serving as a "reference" category. With R, the reference category is always the first level of the factor, following the lexicographical order i.e. in this case the level White. The regression coefficients then represent the difference in average weight between infants of mothers in the category Black versus White (-383.03), and Other versus White (-297.44). It can be verified by calculating these differences in means manually:

```
> grp.means <- with(birthwt, tapply(bwt, race, mean))
> grp.means[2:3] - grp.means[1] # regression model reg.res2
```

To predict the weight of an individual whose mother weighs 60 kg, the command predict() will be employed, knowing that this same command can also provide a confidence interval (on average or for a future observation) with the option interval=.

```
> m <- lm(bwt ~ lwt, data=birthwt)
> d <- data.frame(lwt=60)
> predict(m, newdata=d, interval="confidence")
```

If no value is specified for the option `newdata=`, R will produce an estimate of the predicted value for each statistical unit with the values observed for the cofactor(s) present in the model.

6

Measures of Association
in Epidemiology and
Logistic Regression

This chapter revisits the measures of association typically found in epidemiology or clinical research studies and the concepts discussed partly in Chapter 2. It will cover the risk measures such as the odds ratio in prospective studies or the sensitivity/specificity measures of a diagnostic test. The simple logistic regression model in which we consider a single explanatory variable, numerical or qualitative, is presented in detail: estimation of the parameters of the model, in the case of individual or grouped data, pointwise and interval prediction. Finally, the construction of an ROC curve completes this chapter addressing the case where the response variable is a binary variable.

6.1. Contingency tables and measures of association

6.1.1. *Contingency tables*

As already seen in Chapter 3, the commands `table()` and `xtabs()` allow for building two-dimensional tables or more (this is the case of `xtabs()`, but also of `ftable()`). This recalls the principle of the construction of a table from a matrix for the calculation of frequencies, here with the variables `low` and `race` of the data `birthtw` whose frequencies are manually captured:

```
> tab <- matrix(c(73, 23, 15, 11, 42, 25), nrow=2)
> rownames(tab) <- c(">= 2.5 kg", "< 2.5 kg")
> colnames(tab) <- c("White", "Black", "Other")
```

```
> tab
         White Black Other
>= 2.5 kg   73    15    42
< 2.5 kg    23    11    25
```

When the counts are presented in columns, as in the previous example, it is sufficient to indicate to R how to arrange the table in terms of number of rows or of columns. If the values are entered per line (73, 15, 42 and then 23, 11, 25), the option by.row=TRUE must be added. Naturally, the same results are available by making use of the data set directly:

```
> with(birthwt, table(low, race))
     race
low    White Black Other
  No    73    15    42
  Yes   23    11    25
```

which is equivalent to xtabs(~ low + race, data=birthwt).

6.1.2. *Measures and association tests*

The association measures and tests for two qualitative variables have been presented in sections 3.3 and 3.4. Concerning the odds ratio and the relative risk, particular attention should be given to the orientation of the table (rows and columns) and to the order of the modalities of the binary variables in each dimension (generally, the "positive" events should appear in the first row and in the first column).

6.1.3. *Odds ratio and stratification*

When focusing on the cross-tabulation of two binary variables according to k levels or modalities of a third qualitative variable, considered as a stratification factor, it is possible to calculate k odds ratio and if they prove to be homogeneous it is possible that the strata-specific estimates be combined into a common odds ratio. The test that allows the verification of the homogeneity between strata is the Mantel-Haenszel test and is obtained with the command mantelhaen.test() in R. The data must be arranged in the form of a three-dimensional array, the last dimension representing the stratifying factor levels. There are two approaches to construct such a table: from two-dimensional matrices (matrix()) assembled together to form an array (a generalization of the matrix) or from the command xtabs(), easier to use, in general.

Below is shown how it is possible to proceed with a series of matrices summarizing the frequency table obtained by crossing the modalities of variables `ui` (interuterine pain) and `low` (weight status at birth), for each level of the variable `race` (ethnicity):

```
> t1 <- with(subset(birthwt, race == "White"), table(ui, low))
> t2 <- with(subset(birthwt, race == "Black"), table(ui, low))
> t3 <- with(subset(birthwt, race == "Other"), table(ui, low))
> tab <- array(c(t1, t2, t3), dim=c(2,2,3),
+              dimnames=list(c("ui=0","ui=1"), c(">= 2.5 kg",
                                                 "< 2.5 kg"),
+                            levels(birthwt$race)))
> tab
, , White

      >= 2.5 kg < 2.5 kg
ui=0         65       18
ui=1          8        5

, , Black

      >= 2.5 kg < 2.5 kg
ui=0         14        9
ui=1          1        2

, , Other

      >= 2.5 kg < 2.5 kg
ui=0         37       18
ui=1          5        7
```

Note that, instead of using the `dimnames=` option in the `array()` command, it is also possible to name the rows and columns of the table *a posteriori* by means of `dimnames()`, as shown hereafter:

```
> dimnames(tab) <- list(c("ui=0","ui=1"),
                    c(">= 2.5 kg", "< 2.5 kg"),
                    levels(birthwt$race))
```

Another possibility is presented here, based on the `xtabs()` command, which indicates with a formula the list of variables to be considered, with the stratification variable in the last position. We will be able to verify whether the following command correctly provides the same table as the previous illustration:

```
> xtabs(~ ui + low + race, data=birthwt)
```

```
, , race = White

     low
ui     No Yes
  No   65  18
  Yes   8   5

, , race = Black

     low
ui     No Yes
  No   14   9
  Yes   1   2

, , race = Other

     low
ui     No Yes
  No   37  18
  Yes   5   7
```

Regardless of the table being used, the Mantel-Haenszel test result indicates that the association between the weight status of babies and the presence of intrauterine pain is not independent of the ethnicity of the mothers, suggesting that it is not possible to derive an estimate of the common odds ratio:

```
> mantelhaen.test(tab)

Mantel-Haenszel chi-squared test with continuity correction

data:  tab
Mantel-Haenszel X-squared = 4.2339, df = 1, p-value = 0.03962
alternative hypothesis: true common odds ratio is not equal to 1
95 percent confidence interval:
 1.133437 6.015297
sample estimates:
common odds ratio
        2.611122
```

R also includes the Woolf test to evaluate the heterogeneity between strata through the command woolf_test() from the package vcd:

```
> library(vcd)
> woolf_test(tab)
```

Woolf-test on Homogeneity of Odds Ratios (no 3-Way assoc.)

data: tab
X-squared = 0.093904, df = 2, p-value = 0.9541

6.2. Diagnostic studies

In situations where the main focus is on the predictive or on the discriminating capability of a screening or detection test (ill/not ill), knowing the diagnosis of patients (based on expert judgement or on an already existing test considered as a "gold standard"), most of the measures of interest (sensitivity, specificity, positive and negative predictive values, odds ratio pre- or post-test, etc.) can be calculated from a contingency table.

6.2.1. Sensibility and specificity of a diagnostic test

As discussed in section 3.4, in order to calculate and to correctly interpret most of the measures of association in a 2×2 table, particular care should be given to the orientation of the table (rows and columns) and to the modalities presented in rows and columns. Consider the following data:

```
> dat <- as.table(matrix(c(670,202,74,640), nrow=2, byrow=TRUE))
> colnames(dat) <- c("Dis+","Dis-")
> rownames(dat) <- c("Test+","Test-")
> dat
      Dis+ Dis-
Test+  670  202
Test-   74  640
```

This concerns the results of a validation study of a new diagnostic test achieved with 1,586 patients [SCO 08] taken from the online help of the package epiR. Among the 744 ill patients, 670 were identified as such by this new test. The command as.table() is not essential for the following calculations. It should simply be noted that rows and columns of the table are named addressing its cells by employing the terms of the two variables, rather than by the row or column numbers.

The sensitivity and specificity can be calculated by simple indexation of the table, as in the following manner:

```
> sens <- dat["Test+","Dis+"]/sum(dat[,"Dis+"])
> spec <- dat["Test-","Dis-"]/sum(dat[,"Dis-"])
> round(c(sens=sens, spec=spec), 3)
```

```
sens  spec
0.901 0.760
```

Approximate confidence intervals can be calculated from the normal distribution, using the formula $p \pm 1,96\sqrt{p(1-p)/n}$, where p is the proportion under study. The bounds of the confidence interval would be thus estimated as hereafter:

```
> sens + c(-1, 1) * 1.96 * sqrt(sens * (1-sens)/sum(dat))
[1] 0.8858082 0.9152670
```

6.2.2. Positive and negative predictive values

The calculation of the positive predictive value (PPV) and negative (NPV) can be done using the same principle:

```
> ppv <- dat["Test+","Dis+"]/sum(dat["Test+",])
> npv <- dat["Test-","Dis-"]/sum(dat["Test-",])
> round(c(PPV=ppv, NPV=pnv), 3)
  PPV   NPV
0.768 0.896
```

Since these are simple proportions, it is perfectly possible to test the null hypothesis $H_0 : p = \frac{1}{2}$ by making use of a binomial test. For the PPV, for example, the following results are obtained:

```
> binom.test(dat["Test+","Dis+"], sum(dat["Test+",]))

Exact binomial test

data:  dat["Test+", "Dis+"] and sum(dat["Test+", ])
number of successes=670, number of trials=872, p-value<2.2e-16
alt hypothesis: true probability of success is not equal to 0.5
95 percent confidence interval:
 0.7388926 0.7959784
sample estimates:
probability of success
            0.7683486
```

This type of binomial distribution can also be employed in the calculation of confidence intervals, as illustrated in the following sections.

6.2.3. *Synthesis table of the diagnostic properties of a test*

The command epi.tests() of the package epiR provides the totality of the previous indicators in a single table, with confidence intervals based on the binomial distribution:

```
> epi.tests(dat, conf.level = 0.95)
          Disease +    Disease -      Total
Test +       670          202          872
Test -        74          640          714
Total        744          842         1586

Point estimates and 95 % CIs:
-----------------------------------------------------------
Apparent prevalence                    0.55 (0.52, 0.57)
True prevalence                        0.47 (0.44, 0.49)
Sensitivity                            0.90 (0.88, 0.92)
Specificity                            0.76 (0.73, 0.79)
Positive predictive value              0.77 (0.74, 0.80)
Negative predictive value              0.90 (0.87, 0.92)
Positive likelihood ratio              3.75 (3.32, 4.24)
Negative likelihood ratio              0.13 (0.11, 0.16)
-----------------------------------------------------------
```

As a matter of fact, this command calculates other quantities of interest that are not displayed, but that can be obtained in a relatively simple manner when the result of the call to the command epi.tests() is stored in an auxiliary variable and the option verbose=TRUE is added. To this end, it is necessary to assign the result produced by the previous statement to a variable called, for example, diagstats:

```
> diagstats <- epi.tests(dat, conf.level = 0.95)
```

From this point, the other statistics calculated from the data dat can be accessed by appending our auxiliary variable with the name of the statistics described in the online help (help(epi.tests)). The results described below correspond to the diagnostic OR and to the number of patients to be tested so as to obtain a correct positive result:

```
> diagstats$rval$diag.or
       est      lower     upper
1 28.68611 21.51819 38.24174
> diagstats$rval$nnd
       est    lower     upper
1 1.513701 1.4091 1.648743
```

6.3. Logistic regression

6.3.1. *Estimation of the model's parameters*

In Chapter 5, the main topic of interest was the relationship between the weight of babies at birth, treated as a numeric variable, and the weight of the mothers. The same kind of relationship can be significant to us, but this time we will consider the weight status at birth addressed as a binary variable (low). In fact, the variable low in the data birthwt is simply a dummy variable built from the variable bwt (low = 1 if bwt < 2,500, 0 otherwise).

To build a logistic regression model, we will make use of the command glm() which works exactly in the same way as the command lm() does for linear regression, except that R should be informed about the type of relationship actually connecting the binary response variable to the linear combination of the predictors of interest. In the case of logistic regression, a logit transformation will be considered rather than the identity function used in linear regression:

```
> glm(low ~ lwt, data=birthwt, family=binomial("logit"))

Coefficients:
(Intercept)            lwt
    0.99831        -0.03093

Degrees of Freedom: 188 Total (i.e. Null);   187 Residual
Null Deviance:       234.7
Residual Deviance: 228.7   AIC: 232.7
```

The default link function being the logit function, we can simplify the option family= by simply stating binomial. Similarly, to the case of the ANOVA or that of the regression, the command glm() itself provides only part of the information and it is necessary to couple it with summary() to obtain the table of the regression coefficients:

```
> m <- glm(low ~ lwt, data=birthwt, family=binomial("logit"))
> summary(m)

Deviance Residuals:
    Min       1Q   Median       3Q      Max
-1.0951  -0.9022  -0.8018   1.3609   1.9821

Coefficients:
            Estimate Std. Error z value Pr(>|z|)
```

```
(Intercept)   0.99831    0.78529    1.271    0.2036
lwt          -0.03093    0.01357   -2.279    0.0227 *
---
Signif. codes:  0 '***' 0.001 '**' 0.01 '*' 0.05 '.' 0.1 ' ' 1

(Dispersion parameter for binomial family taken to be 1)

    Null deviance: 234.67  on 188  degrees of freedom
Residual deviance: 228.69  on 187  degrees of freedom
AIC: 232.69

Number of Fisher Scoring iterations: 4
```

The interpretation of the regression coefficient is relatively simple: –0.03 reflects the variation in log-odds for the binary response variable (`low`) when the variable `lwt` varies of one unit.

The same procedure will be carried out when the explanatory variable is a qualitative variable. For example, when the relationship between weight status at birth and intrauterine pain history is to be modelled, we will have the model `low ~ ui`:

```
> table(birthwt$ui)

 No Yes
161  28
> levels(birthwt$ui)
[1] "No"  "Yes"
```

Particular attention should be given to the fact that the first level of the variable `ui` is the response No (no pain), but if this would have not been the case the two levels could be interchanged such that the interpretation of the regression coefficients be achieved with respect to this level. To this end, it would suffice to change the order of the levels by using the command `relevel()` which makes it possible to change the reference level of a qualitative variable:

```
> birthwt$ui <- relevel(birthwt$ui, ref="No")
```

The results of this second model are indicated hereafter:

```
> m2 <- glm(low ~ ui, data=birthwt, family=binomial)
> summary(m2)
```

```
Call:
glm(formula = low ~ ui, family = binomial, data = birthwt)

Deviance Residuals:
    Min      1Q   Median      3Q      Max
-1.1774  -0.8097  -0.8097   1.1774   1.5967

Coefficients:
            Estimate Std. Error z value Pr(>|z|)
(Intercept)  -0.9469     0.1756  -5.392 6.97e-08 ***
uiYes         0.9469     0.4168   2.272   0.0231 *
---
Signif. codes:  0 '***' 0.001 '**' 0.01 '*' 0.05 '.' 0.1 ' ' 1

(Dispersion parameter for binomial family taken to be 1)

    Null deviance: 234.67  on 188  degrees of freedom
Residual deviance: 229.60  on 187  degrees of freedom
AIC: 233.6

Number of Fisher Scoring iterations: 4
```

The odds ratio is obtained simply by taking the exponential of the regression coefficient (slope), but for the sake of simplicity this transformation can be applied to all the coefficients of the model (we will ensure that in the case of the y-intercept ("Intercept"), a simple odds will then be obtained):

```
> exp(coef(m2))
(Intercept)        uiYes
   0.387931     2.577778
```

To display the odds ratio for the variable ui, it suffices to replace the previous expression by exp(coef(m2)["uiYes"]). With the command oddsratio() from the package vcd, it can be verified that the same value is found for the odds ratio:

```
> oddsratio(xtabs(~ ui + low, data=birthwt), log=FALSE)
[1] 2.577778
```

The command confint() can still be employed with the coefficients of the model to obtain their 95% confidence interval, for example:

```
> confint(m2)
Waiting for profiling to be done...
                2.5 %       97.5 %
```

```
(Intercept) -1.3007911 -0.6105528
uiYes        0.1250472  1.7717715
```

Similarly to the ANOVA table derived from a linear regression analysis, it is possible to provide a deviance table for the logistic regression model. This is always carried out with the command anova(), but specifying the type of test that has to be performed by including the option test=, in this case a χ^2 test rather than a test F:

```
> anova(m2, test="Chisq")
Analysis of Deviance Table

Model: binomial, link: logit

Response: low

Terms added sequentially (first to last)


      Df Deviance Resid. Df Resid. Dev Pr(>Chi)
NULL                   188      234.67
ui     1   5.0761      187      229.60  0.02426 *
---
Signif. codes:  0 '***' 0.001 '**' 0.01 '*' 0.05 '.' 0.1 ' ' 1
```

6.3.2. Predictions with confidence intervals

The command predict() will always be used to estimate the values predicted by the model in present or in future observations. In the latter case, it is important not to forget to provide a data frame with the values of the explanatory variable upon which the predictions are to be calculated.

However, unlike linear regression, it is necessary to indicate what kind of forecasts are we willing to calculate: values expressed on the log-odds scale (type="link') or in terms of response probabilities (type="response'):

```
> head(predict(m2, type="link"), n=4)
           85             86             87             88
 2.220446e-16 -9.469277e-01 -9.469277e-01  2.220446e-16
> head(predict(m2, type="response"), n=4)
        85        86        87        88
0.5000000 0.2795031 0.2795031 0.5000000
```

The command `fitted()` can also be employed to obtain the fitted values computed from the sample. In case of the first model with weight of the mothers, the following would thus be used:

```
> xyplot(fitted(m) ~ lwt, data=birthwt)
```

In the graph generated by this command (Figure 6.1), each point represents an individual prediction.

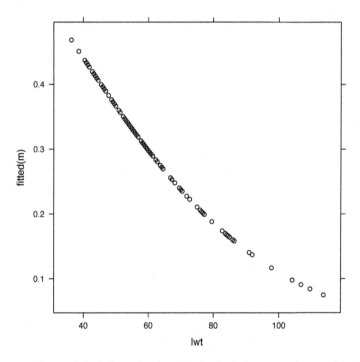

Figure 6.1. *Adjusted values for the logistic regression model*

In the case of forecasting with confidence intervals, `predict()` should be used. However, in contrast with the linear regression case, there is no option `interval=` but it could be very easy to calculate asymptotic intervals from the moment where the standard error of the estimate is available, obtained by including the option `se.fit=`. Given:

```
> pp <- predict(m2, type="link", se.fit=TRUE)
> lwr <- pp$fit - qnorm(0.975) * pp$se.fit
> upr <- pp$fit + qnorm(0.975) * pp$se.fit
> ppd <- data.frame(y=pp$fit, lwr=lwr, upr=upr)
```

```
> head(ppd)
          y          lwr         upr
85  2.220446e-16  -0.7407968   0.7407968
86  -9.469277e-01 -1.2911393  -0.6027161
87  -9.469277e-01 -1.2911393  -0.6027161
88  2.220446e-16  -0.7407968   0.7407968
89  2.220446e-16  -0.7407968   0.7407968
91  -9.469277e-01 -1.2911393  -0.6027161
```

The previous confidence intervals relate to the predictions expressed in logit units. Returning to the probabilities, one can enter the command plogis() and apply it to each of the columns of the table named ppd:

```
> head(sapply(ppd, plogis))
           y         lwr        upr
[1,]  0.5000000  0.3228299  0.6771701
[2,]  0.2795031  0.2156600  0.3537225
[3,]  0.2795031  0.2156600  0.3537225
[4,]  0.5000000  0.3228299  0.6771701
[5,]  0.5000000  0.3228299  0.6771701
[6,]  0.2795031  0.2156600  0.3537225
```

It will be possible to verify that the same results are obtained correctly by performing the computation manually, as hereafter in the case of the upper bound of the 95% CI:

```
> head(exp(pp$fit+1.96*pp$se.fit)/(1+exp(pp$fit+1.96*pp$se.fit)))
      85        86        87        88        89        91
0.677173  0.353724  0.353724  0.677173  0.677173  0.353724
```

6.3.3. The case of grouped data

In the previous example, individual data were available, in other words values of low and lwt for each statistical unit. This is often the case when one of the explanatory variables is continuous. However, in the case of purely qualitative variables, the available data is often presented in form of a two-dimensional contingency table. In the simplest case, this would be a two-dimensional table crossing the two levels of the response variable (ill/not ill) and those of the explanatory variable, that is to say typically a table produced by xtabs(), as for example:

```
> tab <- xtabs(~ ui + low, data=birthwt)
> tab
      low
```

```
ui      No Yes
  No   116  45
  Yes   14  14
```

An alternative representation for this data would consist of indicating, for each level of the explanatory variable the number of positive events (here, low=1) and the total number of events (or observations), that is:

```
> tab2 <- data.frame(ui=c("No","Yes"), n=c(45,14),
                tot=c(45+116, 14+14))
> tab2
    ui  n tot
1   No 45 161
2  Yes 14  28
```

The following two formulations are then equivalent:

```
> glm(n/tot ~ ui, data=tab2, family=binomial, weights=tot)
> glm(cbind(n, tot-n) ~ ui, data=tab2, family=binomial)
```

6.4. ROC curve

There are many R packages allowing the construction of a ROC curve. Here is an illustration with the package ROCR which has the advantage of not having many constraints about data formatting: based on a regression model, in the following example the weight status according to the weight and the ethnicity of the mothers, it suffices to make use of the two commands prediction() and performance() to format the data and to calculate the statistics needed for the ROC analysis. A command plot() ensures that the results be displayed in a graphical manner, depending on the chosen statistics.

In the most conventional case, the tradeoff between the sensitivity (true positives, or tpr) and the complementary of the specificity (false positives, or fpr) are of major interest which can be translated into the following statements:

```
> m3 <- glm(low ~ lwt + race, data=birthwt, family=binomial)
> pred <- prediction(predict(m3, type="response"), birthwt$low)
> perf <- performance(pred, "tpr","fpr")
> plot(perf, ylab="Sensitivity", xlab="1-Specificity")
```

It is necessary to provide both the forecasts of the model (of the "response" type) and the real diagnostic (here, low). The last command allows building the ROC curve (Figure 6.2). A curve that deviates from the first bisector in the upper left quadrant indicates that the observed predictions deviate from predictions randomly achieved.

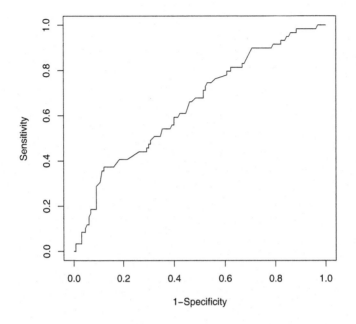

Figure 6.2. *ROC curve for the logistic regression model*

The command `roc.from.table()` from the package `epicalc` allows to work directly from tabulated data (case of the table crossing the results of a diagnostic test with a "gold standard", for example). In the case of a logistic regression, it suffices to provide the variable containing the result of the model, in our case m3. For a table comprising individual data, the command `lroc()` should be entered. The package pROC (http://web.expasy.org/pROC/) essentially provides the same kind of features, with the ability to display smoothed ROC curves and to compare different ROC curves.

6.5. Key points

– In addition to the command `oddsratio()`, the commands `mantelhaen.test()` and `woolf_test()` are also useful for evaluating the conditional independence of the associations between two variables and the homogeneity of odds ratio according to the levels defined by a third qualitative variable.

– The command `epi.tests()` from the package `epiR` provides the main measures allowing for the evaluation of the quality of a diagnostic test.

– The command `glm()` makes it possible to build a logistic regression model; it is employed with `summary()` to obtain the table of the regression coefficients and with `anova()` for the associated deviance table.

6.6. Applications

1) Table 6.1 provides the results of a cohort study aiming to determine, among other things, the advantages of using, as a screening tool, exercise stress test (EST) measures. In such tests, a pass/fail result can be derived during the diagnosis of coronary artery diseases (CAD) [PEP 80].

		CAD		
		Healthy	Ill	Total
EST	Negative	327	208	535
	Positive	115	815	930
	Total	442	1 023	1 465

Table 6.1. *Screening test for physical stress*

It will be assumed that there is no verification bias:

a) Based on this confusion matrix, indicate the following values (with 95% confidence intervals): sensitivity and specificity, positive and negative predictive value.

b) What is the value of the area under the curve for the current data?

The creation of the data table can be done as indicated hereafter, from a table created with the command `matrix()`:

```
> tab <- as.table(matrix(c(815,115,208,327), nrow=2, byrow=TRUE,
                  dimnames=list(EST=c("+","-"),
                               CAD=c("+","-"))))
> tab
```

It should be noted that the table was reorganized so as to correspond to the usual notations, with "positive" events on the first row (typically, the exposure) and the first column (typically, the disease) of the table.

It would be then possible to calculate all the required quantities using very few arithmetic operations with R. However, the package `epiR` contains all the tools needed to respond to epidemiological questions in the 2 x 2 tables. Thus, the command `epi.tests()` provides prevalence, sensitivity/specificity values, as well as positive and negative predictive values:

```
> library(epiR)
> epi.tests(tab)
```

Regarding the area under the curve, an external command can also be employed, roc.from.table(), in the package epicalc:

```
> library(epicalc)
> roc.from.table(tab, graph=FALSE)
```

2) Table 6.2 summarizes the proportion of myocardial infarctions observed in men aged 40 to 59 years and for whom the level of blood pressure and the rate of cholesterol have been measured, considered in the form of ordered classes [EVE 01].

BP	Cholesterol (mg/100 ml)						
	< 200	200 – 209	210 – 219	220 – 244	245 – 259	260 – 284	> 284
< 117	2/53	0/21	0/15	0/20	0/14	1/22	0/11
117 – 126	0/66	2/27	1/25	8/69	0/24	5/22	1/19
127 – 136	2/59	0/34	2/21	2/83	0/33	2/26	4/28
137 – 146	1/65	0/19	0/26	6/81	3/23	2/34	4/23
147 – 156	2/37	0/16	0/6	3/29	2/19	4/16	1/16
157 – 166	1/13	0/10	0/11	1/15	0/11	2/13	4/12
167 – 186	3/21	0/5	0/11	2/27	2/5	6/16	3/14
> 186	1/5	0/1	3/6	1/10	1/7	1/7	1/7

Table 6.2. *Infarction and blood pressure*

The data are available in the file hdis.dat in the form of a table with four columns indicating, respectively, the blood pressure (eight categories, rated 1 to 8), the rate of cholesterol (seven categories, rated 1 to 7), the number of myocardial infarction and the total number of individuals. The association between blood pressure and the likelihood of suffering a myocardial infarction is the point of interest:

a) Calculate the proportions of myocardial infarction for each level of blood pressure and represent them in a table and in graphical form;

b) Express the proportions calculated in (a) in logit form;

c) Based on a logistic regression model, determine whether there is a significant association at the threshold $\alpha = 0.05$ between blood pressure, treated as a quantitative variable by considering the class centers and the likelihood of having a heart attack;

d) Express in logit units the probabilities of infarction predicted by the model for each of the blood pressure levels;

e) Display on the same graph the empirical proportions and the logistic regression curve according to the blood pressure values (class centers).

Importing the data hdis.dat is achieved as hereafter:

```
> bp <- read.table("hdis.dat", header=TRUE)
> str(bp)
```

As it may be verified, there is no label associated with the levels of the variables of interest (bpress for blood pressure, chol for the cholesterol rate). To generate and associate labels, the following commands can be inserted:

```
> blab <- c("<117","117-126","127-136","137-146",
            "147-156","157-166","167-186",">186")
> clab <- c("<200","200-209","210-219","220-244",
            "245-259","260-284",">284")
> bp$bpress <- factor(bp$bpress, labels=blab)
> bp$chol <- factor(bp$chol, labels=clab)
```

The last command converts the original variables into qualitative variables and associates to their modalities the labels defined by blab and clab. To verify that the database is now in the desired form, the following commands are helpful:

```
> str(bp)
> summary(bp)
```

The table of the relative frequencies exposed in the statement can now be reproduced:

```
> round(xtabs(hdis/total ~ bpress + chol, data=bp), 2)
```

Since we are going to focus on the relationship between blood pressure and myocardial infarction only, it is necessary to aggregate data on cholesterol levels. In other words, it is necessary to accumulate the values for each blood pressure level, regardless of the cholesterol levels. The opportunity should also be taken to rename the levels of the variable bpress using the centers of the class intervals:

```
> blab2 <- c(111.5,121.5,131.5,141.5,151.5,161.5,176.5,191.5)
> bp$bpress <- rep(blab2, each=7)
> dfrm <- aggregate(bp[,c("hdis","total")],
                    list(bpress=bp[,"bpress"]), sum)
```

It is the last command, aggregate(), which makes it possible to aggregate the data: it sums all the values (sum) of the variable chol that are not included in the list of variables to be retained for analysis. At the same time, the results are stored in a new database named dfrm. Note that in this case we do not make use of the formula notation but that of the alternative syntax for aggregate() (response variable followed by a list consisting of one or more explanatory variables). An overview of the aggregated data can be obtained by giving head() as the input:

```
> head(dfrm, 5)
```

When a proportion p is available, its value on a scale whose units are logits is given by the relation $\log(p/(1-p))$. Therefrom the following commands can be derived to convert the proportions, calculated like hdis/total in logit units:

```
> logit <- function(x) log(x/(1-x))
> dfrm$prop <- dfrm$hdis/dfrm$total
> dfrm$logit <- logit(dfrm$hdis/dfrm$total)
```

It should be observed that a small function has been defined for converting values x, which here is assumed to be a proportions, in its equivalent $\log(x/(1-x))$. Equally, one could write:

```
> log((dfrm$hdis/dfrm$total)/(1-dfrm$hdis/dfrm$total))
```

The calculated values have been added directly to the data frame under the name prop and logit.

The logistic regression model is written in the following manner:

```
> glm(cbind(hdis,total-hdis) ~ bpress,data=dfrm,family=binomial)
```

The formulation being used, cbind(hdis, total-hdis) ~ bpress, takes into account the fact that we are accessing grouped data and not individual responses. The command glm() with the option family=binomial corresponds to a logistic regression, which, without going much into detail, uses by default the logit function as a canonical link.

The regression coefficients can be displayed by making use of summary():

```
> summary(glm(cbind(hdis, total-hdis) ~ bpress, data=dfrm,
              family=binomial))
```

The previous result includes the essential information to answer the question of the statistical significance of the association between blood pressure and heart attack likelihood: the regression coefficient (on the log-odds scale) is equal to 0.024 and is significant at the usual threshold of 5% (see column Pr(>|z|)).

The likelihood of having a heart attack according to the different levels of blood pressure under consideration is obtained as follows:

```
> glm.res <- glm(cbind(hdis, total-hdis) ~ bpress, data=dfrm,
                 family=binomial)
> predict(glm.res)
```

It should be noted that the intermediate results generated by R have been stored with the name glm.res before using the command predict(). The predictions generated by R are expressed in logit form and it is possible to compare the observed and predicted logits:

```
> cbind(dfrm, logit.predit=predict(glm.res))
```

Virtually all the elements are available to represent the proportions of myocardial infarction observed and predicted graphically according to the level of blood pressure. The predictions of the model expressed in form of proportions and not on the log-odds scale are missing. It may thus be desirable to draw the prediction curve, that is to say the likelihood of having a heart attack according to the blood pressure, without limiting the latter to the eight values observed for the variable bpress. Here follows a possible solution:

```
> dfrm$prop.predit <- predict(glm.res, type="response")
> f <- function(x)
      1/(1+exp(-(coef(glm.res)[1]+coef(glm.res)[2]*x)))
> xyplot(hdis/total ~ bpress, data=dfrm, aspect=1.2, cex=.8,
         xlab="Blood pressure",ylab="Infarction likelihood",
         panel=function(x, y, ...) {
            panel.xyplot(x, y, col="gray30", pch=19, ...)
            panel.curve(f, lty=3, col="gray70")
            panel.points(x, dfrm$prop.predit, col="gray70", ...)
         })
```

3) One case-control investigation focused on the relationship between the consumption of alcohol and that of tobacco and esophageal cancer in humans ("Ille and Villaine" study). The group of cases consisted of 200 patients suffering from esophagus cancer, diagnosed between January 1972 and April 1974. Altogether, 775

control male were selected from the electoral lists. Table 6.3 shows the distribution of the totality of the subjects according to their daily consumption of alcohol, considering that a consumption greater than 80 g is considered to be a risk factor [BRE 80].

	Alcohol consumption (g/day)		
	≥ 80	< 80	Total
Case	96	104	200
Control	109	666	775
Total	205	770	975

Table 6.3. *Alcohol consumption and oesophageal cancer*

a) What is the value of the odds ratio and its 95% confidence interval (Woolf method)? Is it a good estimate of the relative risk?

b) Is the proportion of consumers at risk the same among cases and among controls (consider $\alpha = 0.05$)?

c) Build the logistic regression model making it possible to test the association between alcohol consumption and the status of individuals. Is the regression coefficient significant?

d) Find the value of the observed odds ratio, calculated in (b) and its confidence interval based on the results of the regression analysis.

Since the (individual) raw data are not available, it is necessary to work directly with the frequency table provided in the problem statement:

```
> alcohol <- matrix(c(666,104,109,96), nr=2,
                 dimnames=list(c("Control","Case"),
                               c("<80",">=80")))
> alcohol
```

Regarding the odds ratio, the command oddsratio() from the package vcd will be used:

```
> library(vcd)
> oddsratio(alcohol, log=FALSE)
```

The option log=FALSE ensures that the result returned corresponds correctly to an odds ratio and not to the log of the odds ratio. To obtain an asymptotic confidence interval the following will be employed:

```
> confint(oddsratio(alcohol, log=FALSE))
```

Similarly, `summary(oddsratio(alcohol))` could be used to perform a hypothesis test on the log odds ratio ($H_0 : \log(\theta) = 0$).

The command `prop.test()` can be used to test the hypothesis that the proportion of persons with a daily consumption is ≥ 80 g and is identical in the cases as well as in the controls, indicating the values observed from the cross-tabulation given in the problem statement:

```
> prop.test(c(96,109), c(200,775), correct=FALSE)
```

This test is exactly equivalent to a Z-test to test the difference between two proportions estimated from the data (if no continuity correction is being employed).

With regard to the logistic regression model, the contingency table has to be transformed into a data table in which the two qualitative variables (disease and exposure) are represented properly, that is to say with a data frame. This can be done by using the command `melt()` from the package `reshape2` which makes it possible to transform the data in tabular form into "long" format. At this stage, the levels of the variable disease should be recoded with 0 (control) and 1 (cases), although this is not really necessary and the level 0 should be considered as the reference category (which facilitates the interpretation of the results):

```
> library(reshape2)
> alcohol.df <- melt(alcohol)
> names(alcohol.df) <- c("disease", "exposure", "n")
> levels(alcohol.df$disease) <- c(1,0)
> alcohol.df$disease <- relevel(alcohol.df$disease, "0")
```

The logistic regression model is then written in the following manner:

```
> glm.res <- glm(disease ~ exposure, data=alcohol.df,
                 family=binomial, weights=n)
> summary(glm.res)
```

The result of interest is the row associated with `exposition>=80` as it provides information about the value of the regression coefficient associated to the exposure variable as estimated by R, with its standard error, as well as the value of the test statistic. Here, the regression coefficient is interpreted as the log of the odds ratio. Note that one would obtain exactly the same results by swapping the role of the variables in the previous formulation, `exposure ~ disease`.

The odds ratio calculated above can be found from the regression coefficient associated with the factor of interest (exposure), as well as its 95% confidence interval, in the following manner:

```
> exp(coef(glm.res)[2])
> exp(confint(glm.res)[2,])
```

The second solution consists of considering the number of cases and the total number of individuals:

```
> alcohol2 <- data.frame(expos=c("<80",">=80"), case=c(104,96),
                  total=c(104+666, 96+109))
> summary(glm(cbind(case, total-case) ~ expos, data=alcohol2,
          family=binomial))
```

4) This data comes from a study seeking to establish the prognostic validity of creatine kinase concentration in the body on the prevention of the occurrence of myocardial infarction [RAB 04].

Data are available in the file sck.dat: the first column corresponds to the variable creatine kinase (ck), the second to the presence of the variable disease (pres) and the last to the variable absence of disease (abs).

a) What is the total number of subjects?

b) Compute the relative sick/healthy frequencies and represent their evolution according to creatine kinase values by using a scatterplot (points + segments connecting the points).

c) Based on a logistic regression model which aims at predicting the likelihood of getting sick, calculate the value of CK from which this model predicts that people suffer from the disease considering a threshold of 0.5 (if $P(\text{sick}) \geq 0.5$ then sick=1).

d) Graphically represent the probabilities of being ill predicted by this model as well as the empirical proportions according to the values ck.

e) Establish the corresponding ROC curve and report the value of the area under the curve.

Since the name of the variables is not in the data file, it will be necessary to assign them immediately after data are imported:

```
> sck <- read.table("sck.dat", header=FALSE)
> names(sck) <- c("ck", "pres", "abs")
> summary(sck)
```

The total number of subjects corresponds to the sum of the frequencies for the two variables `pres` and `abs`, that is:

```
> sum(sck[,c("pres","abs")])
```

or, equivalently: `sum(sck$pres) + sum(sck$abs)` (but not `sck$pres + sck$abs`).

To calculate the relative frequencies of these two variables, it is necessary to know the total quantities by variable. These can be obtained by using the command `apply` and operating by columns:

```
> ni <- apply(sck[,c("pres","abs")], 2, sum)
```

Based on this, it suffices to divide each value of the variables `pres` and `abs` by the values calculated previously. The values obtained will be stored in two new variables in the same data table:

```
> sck$pres.prop <- sck$pres/ni[1]
> sck$abs.prop <- sck$abs/ni[2]
```

It is possible to verify that the calculations are correct: the sum of the values for each variable must now be equal to 1:

```
> apply(sck[,c("pres.prop","abs.prop")], 2, sum)
```

Then the proportions obtained are simply represented in a single chart, considering the values of the variable `ck` as the abscissae:

```
> xyplot(pres.prop + abs.prop ~ ck, data=sck, type=c("b", "g"),
      auto.key=TRUE, ylab="Frequency")
```

The instruction `type=c("b", "g")` means that we want the points to be displayed connected by lines (`"b"="o"+"l"`) as well as a grid (`"g"`).

The regression model for grouped data is written as:

```
> glm.res <- glm(cbind(pres,abs) ~ ck, data=sck, family=binomial)
> summary(glm.res)
```

The predictions, expressed in form of probabilities and not on the log-odds scale, are obtained by means of the command predict by specifying the option type="response" as follows:

```
> glm.pred <- predict(glm.res, type="response")
> names(glm.pred) <- sck$ck
```

By considering that the probabilities ≥ 0.5 designate individuals who are ill, the following distribution is thus obtained:

```
> glm.pred[glm.pred >= 0.5]
```

It is concluded that people will be considered as ill, according to this model, for values ck of 80 or more. This is easily verified with a chart where the probabilities predicted based on the values of the variable ck are represented:

```
> sck$sick <- sck$pres/(sck$pres+sck$abs)
> xyplot(glm.pred ~ sck$ck, type="l",
         ylab="Probability", xlab="ck",
         panel=function(...) {
           panel.xyplot(...)
           panel.xyplot(sck$ck, sck$sick, pch=19, col="grey")
         })
```

In a first step, it is necessary to "decompress" the grouped data and to create a table with two columns: the first representing the variable ck and the second representing the presence or the absence of disease. We will use the counts by subgroup previously calculated and available in the variable ni:

```
> sck.expand <- data.frame(ck=c(rep(sck$ck, sck$pres),
                                 rep(sck$ck, sck$abs)),
                           sick=c(rep(1,ni[1]), rep(0,ni[2])))
> table(sck.expand$sick)
> with(sck.expand, tapply(sick, ck, sum))
```

The last two commands are intended to ensure that the same counts are encountered and that the distribution of sick people according to the value of ck is correct. It can also be verified that the same results are obtained regarding the logistic regression and at the same time the values predicted in the previous data table can be added (the same procedure as previously could be followed and the predictions could be replicated, but this is simpler this way):

```
> glm.res2 <- glm(sick ~ ck, data=sck.expand, family=binomial)
> sck.expand$prediction <- ifelse(predict(glm.res2,
                                    type="response")>=0.5,
                                    1, 0)
> with(sck.expand, table(sick, prediction))
```

The last command makes it possible to display a confusion matrix in which the real diagnostics and those predicted by the regression model are crossed. The correct classification rates can be compared when we vary the reference threshold:

```
> classif.tab <- with(sck.expand, table(sick, prediction))
> sum(diag(classif.tab))/sum(classif.tab)
```

To display the ROC curve, we will make use of the package ROCR:

```
> library(ROCR)
> pred <- prediction(predict(glm.res2, type="response"),
                  sck.expand$sick)
> perf <- performance(pred, "tpr","fpr")
> plot(perf, ylab="Sensitivity", xlab="1-Specificity")
> grid()
> abline(0, 1, lty=2)
```

The value of the area under the curve is obtained as follows:

```
> performance(pred, "auc")@"y.values"
```

7

Survival Data Analysis

This final chapter contains an introduction to the modeling of survival data with R. Special attention will be devoted to the representation of censored data in R and to the main commands allowing the manipulation and the summarizing of this type of data. The estimation of the survival function by the Kaplan-Meier's method and the Cox regression model are then discussed in further detail.

In this chapter, the illustrations are achieved based on a different data set than the one used so far. Data are available in the package survival and can be loaded in the following manner:

```
> library(survival)
> data(lung)
```

It concerns data on the survival of patients suffering from lung cancer [LOP 94]. The variables of interest are the following: time (survival time in days), status (status at the point date; 1 = censored data, 2 deceased patient), age (age of the patient) and sex (gender of the patient; 1 = male, 2 = female). In total, there are 228 patients.

The package lattice will not be used in this chapter; on the contrary, the basic graphical functions available in the package survival will be used. On the other hand, there is no real difficulty in adapting the graphics provided by the package survival with the help of lattice.

7.1. Data representation and descriptive statistics

7.1.1. *Data representation format*

Censored data represent a separate type, along with dates and chronological series. Basically, censored data management commands are provided in the package `survival`, which is included with the basic packages of R.

If we can access a series of events coded into 0/1 and of associated temporal data (times, dates, etc.), the command to use is `Surv()`: it allows the time measurements (`time=`) and the event (`event=`) to be specified. It is also possible to specify a time interval whenever it applies (see the option `time2=`). In every case, the status of the observation at the endpoint or end of study must allow for the distinguishing of censored events occurred during the follow-up time. Usually, the data is encoded as 0/1 or 1/2, where 0 (respectively, 1) indicates a censored datum. The status can also be coded as TRUE/FALSE, where `event=TRUE` designates the occurrence of the event (death or other) and `event=FALSE` indicates a (right) censored datum. Here is how to proceed with the `lung` data previously loaded:

```
> st <- with(lung, Surv(time=time, event=status))
> head(st)
[1]   306   455   1010+  210   883   1022+
```

The censored data are indicated by the presence of the suffix +. This is the case of the third observation in the previous example: this individual had not died after 1,010 days of follow-up, but there is no information after then. It can be verified from the raw data by displaying the first six values for the vital status of the patients:

```
> head(subset(lung, select=c(time, status)))
  time status
1  306     2
2  455     2
3 1010     1
4  210     2
5  883     2
6 1022     1
```

7.2. Survival function and Kaplan-Meier curve

7.2.1. *Survival table*

It is possible to construct a survival table and to obtain the probabilities of survival, possibly after stratification according to the levels of a group factor, by making use of the command `survfit()`. By default, `survfit()` displays the survival median and its 95% confidence interval:

```
> s <- survfit(st ~ 1, data=lung)
> s
Call: survfit(formula = st ~ 1, data = lung)

records    n.max n.start    events  median 0.95LCL 0.95UCL
    228      228     228       165     310     285     363
```

The option `conf.type=` eventually makes it possible to modify the type of the confidence intervals calculated for the Kaplan-Meier estimator. By default, a logarithmic transformation is employed. In all cases, the confidence intervals are estimated for each point of the survival function. Additional information is available in the online help, see `help(survfit.formula)`.

The command `summary()` enables that the survival table and the non-parametric estimation of the survival function be generated:

```
> summary(s)
```

By default, R displays the values of the survival function $S(t)$ for the totality of the observed times. However, it may be possible to restrict the calculation and the displaying of the results to a certain time window by including the option `times=`, for example:

```
> summary(s, times=seq(1, 200, by=20))
Call: survfit(formula = st ~ 1, data = lung)
```

time	n.risk	n.event	survival	std.err	lower 95% CI	upper 95% CI
1	228	0	1.000	0.0000	1.000	1.000
21	220	8	0.965	0.0122	0.941	0.989
41	217	3	0.952	0.0142	0.924	0.980
61	211	7	0.921	0.0179	0.887	0.957
81	205	7	0.890	0.0207	0.851	0.932
101	196	6	0.864	0.0227	0.821	0.910
121	189	6	0.837	0.0245	0.791	0.887
141	184	5	0.815	0.0257	0.766	0.867
161	176	8	0.780	0.0275	0.728	0.836
181	159	15	0.713	0.0301	0.656	0.774

or more simply, for a given time:

```
> summary(s, times=300)$surv
[1] 0.5306081
```

As discussed in Chapter 3 for the the χ^2 Pearson test, we take advantage of the fact that R produces one set of results with the command summary() but does not display them all. Thus, one way to access all of the information produced by this command is by employing str(summary(s, times=300)) and by selecting the variable containing the desired result.

To compare two groups, it suffices to add the grouping variable (of the factor-type) to the right of the formula, such as:

```
> lung$sex <- factor(lung$sex, labels=c("Male", "Female"))
> s2 <- survfit(st ~ sex, data=lung)
> s2
Call: survfit(formula = st ~ sex, data = lung)
```

	records	n.max	n.start	events	median	0.95LCL	0.95UCL
sex=Male	138	138	138	112	270	212	310
sex=Female	90	90	90	53	426	348	550

The command summary() then operates based on the same principle and provides an estimate of $S(t)$ for each of the groups defined by the levels of the variable sex:

```
> summary(s2, times=seq(5,20,by=5))
Call: survfit(formula = st ~ sex, data = lung)
```

sex=Male

time	n.risk	n.event	survival	std.err	lower 95% CI	upper 95% CI
5	138	0	1.000	0.0000	1.000	1.000
10	138	0	1.000	0.0000	1.000	1.000
15	132	7	0.949	0.0187	0.913	0.987
20	131	0	0.949	0.0187	0.913	0.987

sex=Female

time	n.risk	n.event	survival	std.err	lower 95% CI	upper 95% CI
5	90	1	0.989	0.011	0.967	1
10	89	0	0.989	0.011	0.967	1
15	89	0	0.989	0.011	0.967	1
20	89	0	0.989	0.011	0.967	1

7.2.2. Kaplan–Meier curve

To represent the survival function in the form of a Kaplan–Meier curve graphically, it is necessary to apply the command plot() to the result produced by survfit() (Figure 7.1). It is often more convenient to save in an auxiliary variable the result of the command `survfit()`, as it has been done in the previous paragraph (variables s and s2):

```
> plot(s, xlab="Time", ylab="Survival probability")
```

The censoring appears in the form of small segments along the survival curve.

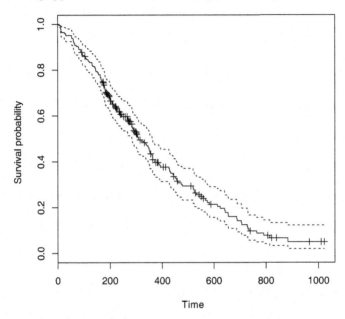

Figure 7.1. *Survival function calculated from the Kaplan–Meier estimator, with 95% confidence interval*

By default, R displays 95% confidence intervals for the survival curve. To remove them, the option `conf.int="none"` must be added. Conversely, when willing to display multiple survival curves, R does not present the confidence intervals, but they can be added by inserting the option `conf.int="both"`. It is also possible to include a legend with the command `legend()` (Figure 7.2):

```
> plot(s2, conf.int="none", col=c("black", "gray"), xlab="Time",
        ylab="Prob. survival")
> legend("bottomleft", legend=levels(lung$sex), lty=1,
        col=c("black", "gray"), bty="n")
```

The option `bty="n"` makes it possible to prevent the display of a frame around the legend. It should be observed that even though the rendering of lines that represent survival curves does not vary (solid line, dotted, etc.), it is nonetheless necessary to indicate an option lty=1 in the command legend() to display lines in the legend box.

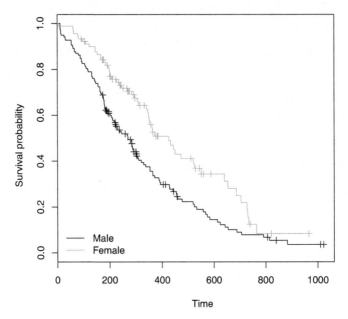

Figure 7.2. *Survival function for each of the levels of the variable sex*

In this case, the graphic commands being used belong to the basic graphical commands in R. However, it is still possible to find the information that is of interest to us, in this case, survival time and probability estimated by the Kaplan–Meier method and to work directly with the command xyplot() of the package lattice.

7.2.3. *Cumulative hazard function*

It is also possible to work with the hazard function $H(t)$ ("hazard rate") which is connected to the survival function by the relation $S(t) = \exp(-H(t))$. One way to estimate this curve consists of employing a transformation like $\hat{H}(t) = -\log \hat{S}(t)$. This is proposed by the command plot() for the objects created with survfit() and by adding the option fun="cumhaz" (Figure 7.3):

```
> plot(s, fun="cumhaz")
```

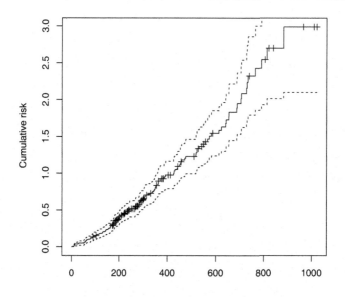

Figure 7.3. *Cumulative risk function*

7.2.4. *Survival function equality test*

To test the null hypothesis of the equality of two or more survival functions, the test log-rank can be utilized, which is available through the command survdiff():

```
> survdiff(st ~ sex, data=lung)
Call:
survdiff(formula = st ~ sex, data = lung)

            N Observed Expected (O-E)^2/E (O-E)^2/V
sex=Male   138      112     91.6      4.55      10.3
sex=Female  90       53     73.4      5.68      10.3

Chisq= 10.3  on 1 degrees of freedom, p= 0.00131
```

If instead, we want to perform a Gehan–Wilcoxon test, it is necessary to specify the option rho=1:

```
> survdiff(st ~ sex, data=lung, rho=1)
Call:
survdiff(formula = st ~ sex, data = lung, rho = 1)

            N Observed Expected (O-E)^2/E (O-E)^2/V
```

```
sex=Male    138     70.4    55.6    3.95    12.7
sex=Female  90      28.7    43.5    5.04    12.7
```

```
Chisq= 12.7  on 1 degrees of freedom, p= 0.000363
```

When it is desirable to give more weight to the first part of the survival function, weights greater than 0 are usually given (rho=), while when is desirable to give more significance to the more distant events in time, negative values will be used for rho=.

7.3. Cox regression

The command to perform a Cox regression is coxph(). Similarly to the linear (Chapter 5) or logistic (Chapter 6) regression case, this command is used to build the regression model while the command summary() will provide the regression coefficients and the associated significance tests:

```
> m <- coxph(st ~ sex, data=lung)
> m
Call:
coxph(formula = st ~ sex, data = lung)


              coef exp(coef) se(coef)     z      p
sexFemale -0.531     0.588    0.167  -3.18 0.0015

Likelihood ratio test=10.6  on 1 df, p=0.00111  n= 228
```

The variable st is naturally the adequate representation for survival data and corresponds to Surv(time, status), in which time denotes the durations and status code for the event (0 = censored data).

It is possible to consider different methods to handle the case of ties with the option method=. By default, R uses Efron's method, but by adding method="breslow" one can access Breslow's method. Stratification factors can also be resorted to by inserting the statement strata() directly in the right member of the R formula specifying the model. For example, the following model is a model stratified on the sex and considering age as the main explanatory variable:

```
> coxph(Surv(time, status) ~ age + strata(sex), data=lung)
Call:
coxph(formula=Surv(time, status) ~ age + strata(sex), data=lung)
```

```
      coef exp(coef) se(coef)    z    p
age 0.0162      1.02  0.00919 1.76 0.078
```

```
Likelihood ratio test=3.18  on 1 df, p=0.0744  n= 228
```

The regression coefficients table is given hereafter:

```
> summary(m)
Call:
coxph(formula = st ~ sex, data = lung)

  n= 228, number of events= 165

              coef exp(coef) se(coef)      z Pr(>|z|)
sexFemale -0.5310    0.5880   0.1672 -3.176  0.00149 **
---
Signif. codes:  0 '***' 0.001 '**' 0.01 '*' 0.05 '.' 0.1 ' ' 1

          exp(coef) exp(-coef) lower .95 upper .95
sexFemale     0.588      1.701    0.4237     0.816

Concordance= 0.579  (se = 0.022 )
Rsquare= 0.046   (max possible= 0.999 )
Likelihood ratio test= 10.63  on 1 df,   p=0.001111
Wald test            = 10.09  on 1 df,   p=0.001491
Score (logrank) test = 10.33  on 1 df,   p=0.001312
```

As in the case of the other regression models discussed in the previous chapters, it is possible to obtain the coefficients by using the command coef() directly.

The Nelson–Aalen estimator can be found from the Cox model in the following manner:

```
> sna <- survfit(coxph(st ~ 1), type="aalen")
> summary(sna)
```

and compare the results with the survival probabilities estimated by the Kaplan–Meier method (section 7.2.3).

```
> h <- -log(sna$surv)
> hest <- data.frame(time=sna$time, d=sna$n.event, n=sna$n.risk,
                     aalen=h, s=sna$surv)
> kmest <- data.frame(time=s$time, s=s$surv)
> psurv <- merge(hest, kmest, by="time")
```

The previous commands allow to build a survival table, having for each time value the number of observed events (d), the number of people at risk (n) and the Nelson–Aalen's estimator (aalen) calculated as the anti-log of the survival probability estimated by means of the Cox model. The command merge() is utilized to assemble the two data tables, the first containing the data calculated with the Cox model (hest) and the second with the data calculated by the actuarial method at the beginning of the chapter (kmest). The join operation of the two tables operation is done based on the variable time. After merging both tables, the variable s.x designates the Nelson–Aalen estimator and s.y the Kaplan–Meier estimator for the survival function. They can be renamed for clarity and the two survival curves displayed as shown in Figure 7.4:

```
> xyplot(s.x + s.y ~ time, data=psurv, type=c("l", "g"),
         ylab="Prob. survival",
         auto.key=list(corner=c(0,0), text=c("Nelson-Aalen",
                                    "Kaplan-Meier"),
              lines=TRUE, points=FALSE))
```

Finally, the command cox.zph() allows the hypothesis of proportional hazards to be tested; two methods are provided: by default, this command displays the result of a χ^2 test for each variable of the model as well as a global test (these tests remain sensitive to the linear tendency at the hazard rate level). It can nonetheless be coupled with a command plot() to visualize the evolution of the Schoenfeld residuals according to time (after transformation) graphically. It is then possible to add an interaction term between a cofactor and the time such as to determine the presence of an interaction between these two variables or construct a stratification factor (see, for example, [FOX 11]).

7.4. Key points

– Survival data are represented in R with the command Surv() in which one indicates the information about time and events (death or other and censoring).

– The Kaplan-Meier estimator is obtained with the command survfit() and it is possible to compare different survival functions through the command survdiff() (log-rank or Wilcoxon test).

– The Cox model works on the same principle as the other regression models and it uses a formula notation to represent the relationship between a response variable of the survival type and explanatory variables. The command summary() makes it possible to display the coefficients of the model.

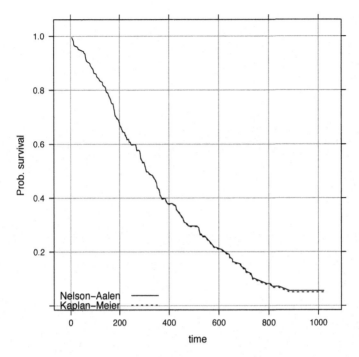

Figure 7.4. *Comparison of the survival functions using the Kaplan–Meier or the Nelson–Aalen estimator*

7.5. Going further

Despite dating from 1999, the manual *A Package for Survival Analysis in S* [THE 99], written by the author of the survival package, provides plenty of additional information. Similarly, the interested reader will find a more thorough development about survival data modeling with R in the book by Frank Harrell [HAR 01].

7.6. Applications

1) In a placebo-controlled trial on biliary cirrhosis, D-penicillamine (DPCA) has been introduced in the active arm with a cohort of 312 patients. In total, 154 patients were randomized in the active arm (treatment variable, rx, 1=Placebo, 2=DPCA). Data such as age, biological data and varied clinical signs including the level of serum bilirubin (bilirub) are available in the file pbc.txt [VIT 05]. The patient's status is stored in the variable status (0=alive, 1=deceased) and the duration of the follow-up (years) represents elapsed time in years since the date of diagnosis.

a) How many deceased individuals can be identified? What proportion of these deaths can be found in the active arm?

b) Display the distribution of the follow-up durations of 312 patients, by distinctively bringing forward the deceased individuals. Calculate the monitoring median time (in years) for each of the two treatment groups. How many positive events beyond 10.5 years are there and what is the gender of these patients?

c) The 19 patients, whose number (variable number) appears amongst the following list have undergone a transplant during the follow-up period:

5 105 111 120 125 158 183 241 246 247 254 263 264 265 274 288 291 295 297 345 361 362 375 380 383

Indicate their average age, the distribution according to sex and the median duration of the follow-up (in days) until transplant.

d) Display a table summarizing the distribution of risk events according to time, with the associated survival value.

e) Display the Kaplan-Meier curve with a 95% confidence interval, without considering the type of treatment.

f) Calculate the survival median and its 95% confidence interval for each group of subjects and display the corresponding survival curves.

g) Perform a log-rank test considering as predictor the factor rx. Compare with a Wilcoxon test.

h) carry out a log-rank test on the factor of interest (rx) by stratifying the age. Three age groups will be considered: 40 years old or less, between 40 and 55 years age inclusive and more than 55 years old.

i) Recover the results of exercise 1.g with a Cox regression.

Loading the data raises no specific difficulties:

```
> pb <- read.table("data/pbc.txt", header=TRUE)
> names(pb)[1:20]
```

To facilitate the reading of the results returned by R, we can recode the variables treatment (rx) and sex (sex) as categorical variables:

```
> pb$rx <- factor(pb$rx, labels=c("Placebo", "DPCA"))
> table(pb$rx)
> pb$sex <- factor(pb$sex, labels=c("M","F"))
```

The number or the proportion of patients who died can be obtained directly from a frequency table of the variable `status`:

```
> prop.table(table(pb$status))
```

The proportion of deaths by treatment group can be obtained by crossing the two variables `status` and `rx` and by calculating the relative frequencies on each row:

```
> prop.table(with(pb, table(status, rx)), 1)
```

To display the distribution of the follow-up duration, a simple scatter plot can be used in which the coordinates of the points to be displayed are defined by the observation number and by the duration of the follow-up from the diagnostic:

```
> xyplot(number ~ years, data=pb, pch=pb$status, cex=.8)
```

The follow-up median time by treatment group is obtained from a command like `tapply()` and considering the variable `rx` as a classification variable:

```
> with(pb, tapply(years, rx, median))
```

The number of patients who died beyond 10.5 years is obtained as follows:

```
> with(pb, table(status[years > 10.5]))
> with(pb, table(sex[years > 10.5 & status == 1]))
```

that is four patients, including two female patients. Note that the last command could also be written:

```
> subset(pb, years > 10.5 & status == 1, sex)
```

which has the advantage of providing the observation numbers.

Transplanted patients are normally among the individuals alive at the transition date. This can be easily verified by making use of a single variable frequency analysis of the `status` of these patients:

```
> idx <- c(5,105,111,120,125,158,183,241,246,247,254,263,264,265,
           274,288,291,295,297,345,361,362,375,380,383)
> table(pb$status[pb$number %in% idx])
```

As a first step and to calculate the required quantities (average age, sex and follow-up duration), the initial data table can be reduced to these patients only:

```
> pb.transp <- subset(pb, number %in% idx, c(age, sex, years))
```

Next, it simply suffices to apply the commands `mean()`, `table()` and `median()` to the variables selected for these 19 patients:

```
> mean(pb.transp$age)
> table(pb.transp$sex)
> median(pb.transp$years * 365)
```

It is necessary to use the command `Surv()` from the package `survival` to obtain a table of the events with the associated survival; an example of its utilization follows:

```
> library(survival)
> head(with(pb, Surv(time=years, event=status)))
```

Censored data appear appended with a +. A table of the events according to time can be constructed by inserting the command `survfit()`. It should be noted that when no group or stratification factor is being considered, one should employ the slightly peculiar syntax `Surv(time, event) ~ 1`:

```
> s <- survfit(Surv(years, status) ~ 1, data=pb)
> summary(s)
```

Once the survival table has been built, it is very simple to display the survival estimator in the form of a Kaplan–Meier curve by making use of the command `plot()`:

```
> plot(s)
```

The median survival for each treatment arm can be obtained with the command `survdiff()`. Firstly, we redefine the survival model to include the predictor of interest:

```
> s <- survfit(Surv(years, status) ~ rx, data=pb)
```

To obtain the survival median, it suffices to show the previous result by simply typing the name of the variables in which the result returned by the command `survfit` was stored. We will use the command plot(s) to obtain the corresponding survival curves.

The log-rank test is achieved with the command survdiff(), with the same notations as in the case of the calculation of the survival median:

```
> survdiff(Surv(years, status) ~ rx, data=pb)
```

The option rho= allows the type of the test performed to be controlled. By adding rho=1, a Wilcoxon test is carried out instead of the conventional log-rank:

```
> survdiff(Surv(years, status) ~ rx, data=pb, rho=1)
```

To achieve a stratified analysis based on the age, it is necessary to recode this numeric variable into a qualitative variable, which can be performed by using the command cut():

```
> agec <- cut(pb$age, c(26, 40, 55, 79), include.lowest=TRUE)
```

The log-rank test can again be carried out including this additional stratification factor and by slightly adapting the previous model:

```
> survdiff(Surv(years, status) ~ rx + strata(agec), data=pb)
```

Finally, for the Cox model, the command coxph() has to be used.

```
> summary(coxph(Surv(years, status) ~ rx + strata(agec), data=pb))
```

2) In Table 7.1 one can find the durations of remission in weeks in a trial comparing 6-MP to a placebo [FRE 63]. The censored observations (patients without relapse, according to the latest information) are indicated with an asterisk.

a) Build a table indicating the number of exposed people and the number of relapses according to time for the group 6-MP.

b) Display the survival curve in the group 6-MP, as well as its 95% confidence interval (Kaplan-Meier estimator).

c) Display the cumulative hazard function (Nelson estimator) for the same group.

d) Calculate the survival median of the group 6-MP.

e) Calculate the expected survival for one patient of the group 6-MP at fifteen weeks.

Placebo	1 1 2 2 3 4 4 5 5 8 8 8 8 11 11 12 12 15 17 22 23
6-MP	6 6 6 6* 7 9* 10 10* 11* 13 16 17* 19* 20* 22 23 25* 32* 32* 34* 35*

Table 7.1. *6-MP data*

In the first stage, a table of the raw data has to be built, in which the treatment, the survival duration and censoring will be presented (0 = alive/1 = deceased):

```
> placebo.time <- c(1,1,2,2,3,4,4,5,5,8,8,8,8,11,11,12,12,15,17,
                 22,23)
> placebo.status <- rep(1, length(placebo.time))
> mp.time <- c(6,6,6,6,7,9,10,10,11,13,16,17,19,20,22,23,25,32,
            32,34,35)
> mp.status <- c(1,1,1,0,1,0,1,0,0,1,1,0,0,0,1,1,0,0,0,0,0)
> mp <- data.frame(tx=rep(c("Placebo","6-MP"), c(21,21)),
                time=c(placebo.time, mp.time),
                status=c(placebo.status, mp.status))
> summary(mp)
```

It can be verified that the events have been correctly registered in this new data table by verifying the format of the survival data in R:

```
> with(mp, Surv(time, status))
```

The table of events in the group 6-MP can be constructed by means of the command `survfit()`. It would also be possible to restrict the calculations to the only observations of the group 6-MP by including the option `subset()`:

```
> s <- survfit(Surv(time, status) ~ tx, data=mp)
> summary(s)
```

The same result will be utilized, contained in the variable s to display the corresponding survival curve. By default, when there is only one group R automatically displays the 95% confidence interval:

```
> plot(survfit(Surv(time, status) ~ tx, data=mp,
            subset=tx=="6-MP"))
```

To display the cumulated hazard function, the option `fun="cumhaz"` will be added to the previous graphical command that is by considering the two groups:

```
> plot(s, fun="cumhaz")
```

3) In a randomized trial, the aim was to compare two treatments for prostate cancer. The patients took each day, by oral route, either 1 mg of diethylstilbestrol (DES, active arm) or a placebo and the survival time is measured in months [COL 94]. The interesting challenge lies in determining if the survival is different between the two groups of patients. Furthermore, the other variables available in the data file `prostate.dat` will be overlooked.

a) Calculate the survival median for the totality of the patients and per treatment group.

b) What is the difference between the survival proportions in the two groups at 50 months?

c) Display the survival curves for the two groups of patients.

d) Perform a log-rank test to verify the hypothesis according to which the DES treatment has a positive effect on the survival of patients.

To import the data, we will use the command `read.table()`. During the prior inspection of the data, the variable `Treatment` will be recoded into a qualitative variable:

```
> prostate <- read.table("prostate.dat", header=TRUE)
> str(prostate)
> head(prostate)
> prostate$Treatment <- factor(prostate$Treatment)
> table(prostate$Status)
```

It is not really necessary to recode `Status` into a categorical variable since this variable is going to be utilized jointly with `Time` in the survival analysis. To transform the data into survival data, including censoring, the first step necessarily involves the command `Surv()` from the package `survival`:

```
> library(survival)
> with(prostate, Surv(time=Time, event=Status))
```

The survival median is obtained by means of `survfit()`, which is a general command that allows manipulating the survival data with the Kaplan–Meier method or the Cox model. If treatment groups are not taken into account (and more generally any cofactor), then it will be employed as follows:

```
> survfit(Surv(Time, Status) ~ 1, data=prostate)
```

On the other hand, to account for the treatment factor in the analysis, we will make use of:

```
> survfit(Surv(Time, Status) ~ Treatment, data=prostate)
```

To display the survival curves corresponding to the two treatment groups, the procedure will be as follows:

```
> plot(survfit(Surv(Time, Status) ~ Treatment, data=prostate))
```

The log-rank test can be obtained in two different ways. The simplest one consists of entering command survdiff() for comparing two survival curves based on the Kaplan–Meier estimator:

```
> survdiff(Surv(time=Time, event=Status) ~ Treatment,
        data=prostate)
```

The alternative solution naturally consists of using a Cox regression by inserting the command coxph():

```
> summary(coxph(Surv(time=Time, event=Status) ~ Treatment,
        data=prostate))
```

Appendices

Appendix 1

Introduction to RStudio

This chapter provides a brief overview of the features of RStudio. The reader is invited to refer to the books dedicated to RStudio [VER 11, GAN 13] for more information.

A1.1. Getting started

By default, the RStudio graphical user interface displays four panels arranged in the same window. It is possible to maximize the panel containing the R console in order to display the maximum information regarding the output generated by R.

RStudio also offers a file browser that, despite being simplified, allows the user to change the working directory (*More > Set As Working Directory*), similarly to the command `setwd()`.

A1.2. Writing R scripts

While launching the application for the first time, the script editor is not visible and users have to create a new R script to access it. To create a new R script, it suffices to navigate to the menu *File > New File > R Script* or to click on the corresponding button in the toolbar. Every new script is saved automatically, even if the application freezes and has to be restarted. The extension `.R` is added automatically to any new script that is saved.

The script editor consists of a main window with a toolbar. The keywords of the R language are highlighted in a different color and the opening and closing parentheses are automatically matched. An automatic color-coding system also facilitates the organization of the R code.

From within the script editor, it is possible to send any command or several lines of code to the R console; click the *Run* button by selecting the line(s) to be evaluated, or simply use the corresponding keyboard shortcut. The *Source* button makes it possible to send all of the R script to the console. All commands will be executed, but in a silent manner; results will not necessarily be displayed unless commands like `print()` are included inside the script itself. On the contrary, the *Source with echo* button will enable the display of the intermediate results generated by R.

The script editor provides other interesting features. For example, by pressing the *TAB* key, RStudio will complete the names of the commands predefined in R automatically, as well as the name of the variables or functions defined in the R script. The key *F1* can be used to access online help on the command upon which the cursor is located. It is also possible to automatically comment on one or more rows from within the toolbar or with a dedicated keyboard shortcut.

A dedicated dialog box (*Tools > Import Dataset > From Text File*) allows data files to be imported and a preview of the data is available once they are imported into R, including the options specified in the dialog box (field separator, decimal separator, etc.). These are the options that can be found in the command `read.table()`.

A1.3. Management of the R session

RStudio provides very useful functionality for management of the workspace. The workspace can be viewed directly in a dedicated panel (*Environment*). It essentially provides the same information as the command `str()`, namely the variable type (vector, matrix and data frame), the number of elements it contains, and an overview of the first values. In case where the variable is a data frame, an icon in the shape of a spreadsheet appears on its right; double clicking on it visualizes the data frame directly into a spreadsheet in read-only mode. This spreadsheet enables filtering and sorting the observations by columns.

All R commands evaluated in the console are saved in a history accessible through a specific panel (*History*). Each of these commands can be entered again in the console, or even in the script editor if an R script has been created, by selecting the command(s) of interest and clicking on the corresponding button in the mini toolbar.

RStudio offers a specific panel to install packages (*Packages*), where it is possible to specify one or more packages to be downloaded and installed from the CRAN site, which advantageously replaces the use of `install.packages()`. It is also possible to load packages by clicking on a checkbox or to update all of the packages already installed.

Numerous other features are available in RStudio. We shall mention in particular the ability to manage projects comprising several R script files, with a version management system (SVN or Git), as well as the automatic creation of reports making use of the R Markdown language (http://rmarkdown.rstudio.com).

Appendix 2

Graphs with
the Lattice Package

A2.1. Why the lattice package?

There are three main graphical systems in R:

– the base graphics commands, in the graphics package, includes most of the tools necessary for uni- and multivariate representation of numerical and categorical data. Typical commands include plot(), hist(), barplot(), dotchart() and qqplot(). For illustrations, see:

```
> demo(graphics)
```

– the lattice package (http://lmdvr.r-forge.r-project.org/) provides commands equivalent to the base commands, with a unified data entry interface, based on the use of R formulas, more uniform options by default and, in most cases, largely satisfactory print-ready graphics, as well as the automatic construction of so-called "trellis" graphs;

– the ggplot2 package (http://ggplot2.org), which is based on the grammar of graphics [WIL 05], stands out from the two other graphic systems by making use of different commands and syntax, while offering an unparalleled degree of flexibility for creating statistical graphics. The two latter graphic systems are based on the same grid package.

Throughout this book, we have favored the lattice package for several reasons. The relationships between the variables are specified by means of R formulas (and of an option data=). These are formulas that are also utilized in statistical models (ANOVA and regression among others) and some digital summary commands (aggregate, and summary.formula or summarize of the Hmisc package, see

Appendix). The default options (axes annotation, size of font and graphical symbols), are generally satisfactory. For a quick visualization, there are interesting options that avoid having to combine several graphics commands. For example:

```
> xyplot(y ~ x, data=d, type=c("p", "g", "r"))
```

allows displaying a scatterplot (y according to x) with a regression line, whereas by employing the base graphics a series of three commands would have to be used:

```
> with(d, plot(x, y))  # or plot(y ~ x, data=d)
> abline(lm(y ~ x, data=d))
> grid()
```

The following paragraphs provide an overview of the main options. In most cases, only the main keywords are mentioned so that the reader can find his or her way in the online help (which is rather complex for this package).

A2.2. Principal types of graphics

In the following, the principal graphics commands for uni- and bivariate distributions are presented, with eventually a conditioning of the graphical results on the levels of one or more categorical variables. It is also possible to condition the relationship between two variables by a third numeric variable but this case will not be addressed (see the concept of shingle). In each case, the equivalent base command is indicated in parentheses. For the sake of simplicity, the name of the data frame will systematically be omitted and no conditioning or stratification factor will explicitly be considered. As illustrative variables, x, y (numeric variables) and z (categorical variable) will be used.

1) histogram(~ x) (hist(x)) displays a proportion, a count (type="count") or a density (type="density") histogram. Possibility of conditioning with the formula operator | z.

2) densityplot(~ x) (plot(density(x))) displays a non-parametric density curve. Possibility of conditioning with the formula operator | z and groups=z.

3) stripplot(~ x) or stripplot(z ~ x) (plot(z ~ x)) displays a univariate dispersion diagram, eventually conditioned on a variable appearing on the left of the operator ~ in the formula. Possibility of additional conditioning with the formula operator | z and groups=z.

4) bwplot(~ x). (boxplot(x)) Displays a diagram in the form of a boxplot. The conditioning is done directly in the formula, for example bwplot(z ~ x) (bars horizontally oriented) or bwplot(x ~ z) (bars vertically oriented). The formula operator | z can also be employed.

5) `xyplot(y ~ x)` (`plot(y ~ x)` or `plot(x, y)`) displays a scatterplot. Conditioning possibility with the formula operator | z and `groups=z`.

6) `barchart(xtabs(~ z))`   (`barplot(table(x))`)   displays   a   bar chart. Conditioning possibility with the formula operator | z, or by directly providing a two-entry frequency or proportions table, for instance `barchart(xtabs(~ z1 + z2), stack=FALSE)`.

7) `dotplot(xtabs(~ z))` (`dotchart(table(x))`, or `dotchart(x)` if x is already a series of aggregated data, for example means or proportions). Possibility of conditioning with the formula operator | z and `groups=z`. Note: `dotplot(z ~ x)` is essentially equivalent to `stripplot(z ~ x)`.

8) `splom(data.frame(x, y))` or `splom(d)`, with d a `data.frame` containing numerical variables. Possibility of conditioning with the formula operator | z and `groups=z`.

9) `qqmath(~ x)` displays a quantile-quantile plot. Conditioning possibility with the formula operator | z and `groups=z`.

In general, when an additional classification factor is added, the conditioning can be achieved in two ways. If the variable is placed in the R formula with the operator | z, then R will produce as many graphics as there are levels for the variable z (see Figure A2.1):

```
> xyplot(bwt ~ lwt | ui, data=birthwt)
```

The other option consists of specifying an option `groups=` that will allow data conditioned with a classification factor to be overlayed on the same graph data. In the previous example, this would entail displaying all of the weight observations in the same scatterplot, making use of different symbols according to the value of `ui`.

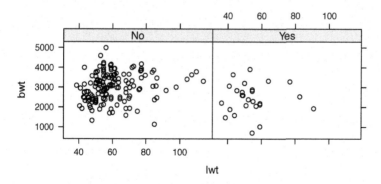

Figure A2.1. *Scatterplot conditioned on a qualitative variable*

In some situations, it is advantageous to transform the data table by employing the command `melt()` of the package `reshape2`. This is, particularly the case when it is desirable to display in the same chart boxplots for different numeric variables of a data frame, but without considering any classification factor. Concerning the birth weight data, one might want to display the distribution of the variables `bwt`, `lwt` and `age`. A command such as:

```
> bwplot(~ bwt + lwt + age, data=birthwt)
```

will not work, because in this case R will display the distribution of the sum of the three variables (for each observation). On the contrary, and subject to the problem of the normalization of the variables that can be solved previously with the command `scale()`, an instruction of the following form will be used (Figure A2.2):

```
> library(reshape2)
> d <- melt(as.data.frame(sapply(birthwt[,c("bwt","lwt","age")],
                          scale)))
> bwplot(variable ~ value, data=d)
```

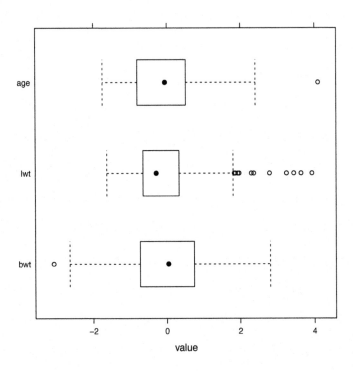

Figure A2.2. *Representation of several variables in the form of boxplots*

This notation makes it possible to display the distribution of each variable in separate histograms as illustrated in Figure A2.3. The significance of transforming the initial data table to reexpress the relationships between variables can therefore be seen:

```
> histogram(~ value | variable, data=d)
```

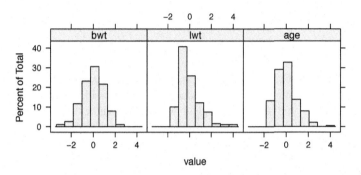

Figure A2.3. *Histograms for several variables*

A2.3. General options for the customization of charts

A2.3.1. *Graphics elements type and annotation*

Concerning the annotation of the axes or the customization of the graphical elements (points, lines, etc.), it is possible to use the same options as for the base graphics, namely:

– main=, xlab= and ylab= to modify the title, and the name of the x and y axes, respectively. The option is a string of characters, for instance "My title";

– pch= and cex=, to modify the type (1 = circle, 2 = square, etc.) and the relative size of the symbols (the values < 1 designate small sized symbols);

– lty= and lwd=, to modify the type (1 = solid line, 2 = dashes, etc.) and the relative size of the lines;

– col= to change the color of the points or of the lines.

The user can refer to the online help for further information: help(par).

For example, the command will display a scatterplot in which the babies' weight is represented on the y-axis (vertical axis, labelled "Babies' weight") according to the mothers' weight, represented on the x-axis (horizontal axis, labelled "Mothers' weight"). The observations will be represented by squares (pch=2), of small size (cex=.6, or 60% of the default size) and with the color green:

```
> xyplot(bwt ~ lwt, data=birthwt, pch=2, cex=.6, col="green",
          xlab="Mothers' weight (kg)", ylab="Babies' weight (g)")
```

A2.3.2. *Automatic management of the graphic elements through the usage of themes*

The options described in the previous section are convenient for simple graphics, but in the case where there is a classification factor, it is more convenient to make use of a "theme" that allows the automatic management of the symbols associated with each level of the classification factor. It also makes it possible to correctly indicate the graphics options applied in the legend be correctly indicated. To this end, the command simpleTheme() of the package latticeExtra (to be installed beforehand) will be employed in the following manner:

```
> xyplot(bwt ~ lwt, data=birthwt, groups=ui,
          xlab="Mothers' weight (kg)",
          ylab="Babies' weight (g)",
          par.settings=simpleTheme(pch=c(1,3)),
          auto.key=TRUE)
```

The overall color theme can also be changed by inserting the command trellis.par.set(). Here follows an example in which we use a theme of a specific color with transparent backgrounds for the conditional graphs:

```
> library(latticeExtra)
> trellis.par.set(custom.theme.2())
> trellis.par.set("strip.background"=list(col="transparent"))
```

To obtain a black and white or grayscale theme, the following instructions can be employed:

```
ltheme <- canonical.theme(color=FALSE)
ltheme$strip.background$col <- "transparent"
lattice.options(default.theme=ltheme)
```

It can be observed that these modifications take effect on the same graphic "device": when closing the graphical window opened by default by R, they will no longer apply.

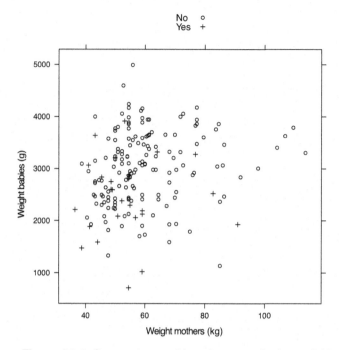

Figure A2.4. *Scatterplot conditioned on a qualitative variable*

A2.3.3. *Graphic legend*

The option auto.key= makes it possible to generate a legend corresponding to the graphical elements present in the figure. When using simpleTheme(), there is no need to worry about the color of the points or about the line type, it is sufficient to manage the placement of the legend.

It is possible to indicate in which margin the legend will be displayed (by default, on the right) or to indicate relative coordinates (between 0 and 1 for each axis). For example, auto.key=list(space="top") instructs R to place the legend in the top margin of the plot. The layout of the different symbols in the legend can also be changed by adding columns=2: the elements will thus be arranged horizontally in two locations (if there are more than two levels for the conditioning factor, the legend items will be stacked on the same two locations). Regarding the positioning in relative coordinates, space= will be replaced by corner= further specifying a list for the coordinates (x, y) of the left upper part of the frame of the legend. There are several other options for the legend option.

Finally, it is possible to add a legend title with a statement such as `title=`. By recalling one of the previous examples, the following commands will generate the graph reproduced in Figure A2.5:

```
> xyplot(bwt ~ lwt, data=birthwt, groups=smoke,
         par.settings=simpleTheme(pch=c(1,3),
                                  col=c("gray50", "black")),
         auto.key=list(corner=c(0.7,0.95), title="Smoking",
                       cex=.8, cex.title=0.8, border=TRUE))
```

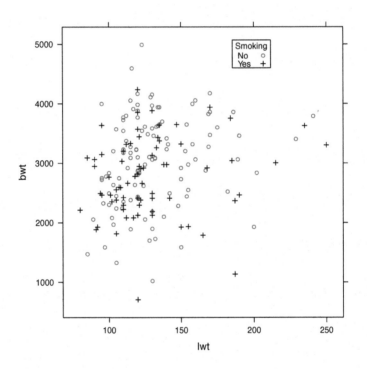

Figure A2.5. *Legend specification in a lattice graph*

A2.3.4. *Combination of graphic elements*

In the case of scatterplots produced with `xyplot()`, the option `type=` is helpful to combine different graphic elements: `"p"` displays the data in the form of points; `"l"` allows connecting each observation by lines; `"b"` displays both points and lines; `"smooth"` will display a loess curve, eventually for each classification factor level indicated in an `groups=` option; `"r"` displays the regression line (eventually several regression lines according to the conditioning option `groups=`); `"g"` enables the displaying of a grid aligned on the intervals defined for both axes.

A2.3.5. *Spatial organization of several graphs*

The option `layout=` makes it possible to specify the arrangement of multiple charts in the presence of a conditioning variable. By using this option, one indicates the number of columns and of lines over which the graphics have to be arranged. Here is an example of the usage (Figure A2.6):

```
> xyplot(bwt ~ lwt | race, data=birthwt, type=c("p", "smooth"),
    layout=c(1,3))
```

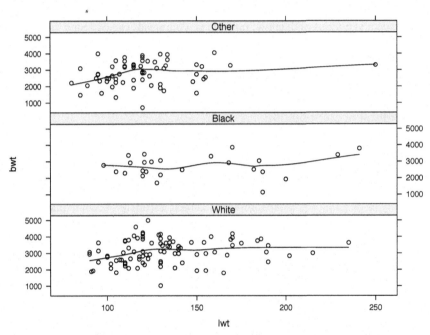

Figure A2.6. *Redisposition of the lattice subgraphs*

The command `grid.arrange()` of the package `grid` allows that different graphics be combined on the same window.

A2.4. Specific options

Some graphics commands have options that are specific to them. For example, the commands `stripplot()` and `xyplot()` both offer the possibility to add a slight offset (horizontal and/or vertical) of the points to avoid the overlapping problems in the case of large samples. With `stripplot()`, for instance one would utilize Figure A2.7:

```
> stripplot(~ bwt, data=birthwt, jitter.data=TRUE, factor=2)
```

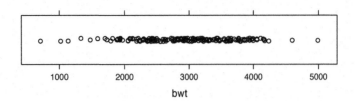

Figure A2.7. *Univariate scatterplot*

Most specific options of the principal graphic commands can be found by accesing the help for the command `panel.*`, where `*` usually refers to the principal graphic command. For example, `help(panel.xyplot)` provides a list of the possible options for representing the relationship between two variables in the form of a scatterplot.

A2.5. Graphical composition

A command such as:

```
> xyplot(bwt ~ lwt, data=birthwt, type=c("p", "g", "r"))
```

can actually be decomposed in the following manner: a scatterplot is built that displays points ("p") to represent the pairs of values observed for the variables `bwt` and `lwt`, then a grid aligned with the axes' coordinates is superimposed, and finally a regression line corresponding to the model `lm(bwt ~ lwt)` is plotted. It is possible to reproduce these steps by using the following instructions:

```
> xyplot(bwt ~ lwt, data=birthwt,
         panel=function(x, y, ...) {
           panel.xyplot(x, y, ...)
           panel.grid(...)
           panel.lmline(x, y, ...)
         })
```

The instruction `panel=function(x, y, ...) { ... }` is responsible for displaying various graphic elements, according to the data `x` and `y` corresponding to the variables `lwt` and `btw`, respectively. Next, the following graphic elements are added: points whose coordinates are defined by `x` and `y` (therefore, the values taken by `lwt` and `btw`) with `panel.xyplot()`, a grid aligned on the axis coordinates being displayed with `panel.grid()` and finally a regression line with the command `panel.lmline()`.

There are more complex cases than the previous example, notably when the graphical command includes a conditioning variable producing different graphs (operator |) or different types of points or lines (options groups=). These cases can be managed by adding an option subscripts= or groups= in the declaration of the function panel. See also the commands packet.number() as well as panel.number() in the online help.

A2.6. Going further

The definitive reference on the package lattice remains the book by Deepayan Sarkar [SAR 08]. Useful information can also be found on the associated website, particularly the R commands to generate different types of graphics: see http://lmdvr. r-forge.r-project.org/figures/figures.html.

The package latticeExtra (http://latticeextra.r-forge.r-project.org) offers numerous extensions to the lattice commands, for multivariate charts and for the customization (theme) of the graphics.

Appendix 3

The Hmisc and rms Packages

The packages Hmisc and rms provide a set of additional functionalities for the management of data, the generation of descriptive summary tables and graphs and for modeling numerical and qualitative variables (linear regression, logistic and analysis of survival data).

A3.1. The Hmisc essential commands

A3.1.1. *Data management*

Regarding data management, Hmisc offers several interesting features. This package provides commands enabling the import of CSV (csv.get()), SAS (sas.get()), SPSS (spss.get()) or Stata (stata.get) files. These commands carry out the conversion of variable names in lowercase and allow recoding dates, if necessary. The last two commands are, in fact, based on the commands of the package foreign. Before using this package, it has to be loaded with the command library(Hmisc).

The command contents() provides a summary of the mode of representation of the variables and of the presence of missing values:

```
> birthwt$age[5] <- NA
> birthwt$ftv[sample(1:nrow(birthwt), 5)] <- NA
> contents(birthwt)

Data frame:birthwt 189 observations and 12 variables    Maximum # NAs:5
```

```
        Levels   Class Storage NAs
low        2           integer  0
age                    integer  1
lwt                    double   0
race       3           integer  0
smoke      2           integer  0
ptl                    integer  0
ht         2           integer  0
ui         2           integer  0
ftv                    integer  5
bwt                    integer  0
ftv2       3           integer  0
ftv3       3 ordered integer  0

+--------+-----------------+
|Variable|Levels           |
+--------+-----------------+
|  low   |No,Yes           |
|  smoke |                 |
|  ht    |                 |
|  ui    |                 |
+--------+-----------------+
|  race  |White,Black,Other|
+--------+-----------------+
|  ftv2  |0,1,2+           |
|  ftv3  |                 |
+--------+-----------------+
```

We have seen that it was possible to associate labels to the modalities of a qualitative variable with the option levels= of the command factor or directly with the command levels(). It is also possible to associate labels to variables, as well as units of measurement, where applicable, by means of the commands label() and units():

```
> label(birthwt$age) <- "Mother's age"
> units(birthwt$age) <- "years"
> label(birthwt$bwt) <- "Baby's weight at birth"
> units(birthwt$bwt) <- "grams"
```

The command label() can also be employed to provide information on a data frame, for example:

```
> label(birthwt, self=TRUE) <- "Hosmer and Lemeshow data
                      on weights at birth."
```

These labels can then be automatically assigned in the tables or in the charts produced by Hmisc. The command describe provides a descriptive summary of the set of variables of a data frame or of a list of variables, similarly to a data dictionary or "codebook":

```
> describe(subset(birthwt, select=c(age, race, low)))
subset(birthwt, select = c(age, race, low))

 4  Variables         189  Observations
------------------------------------------------------------------
age : Mother's age [years]
      n missing  unique   Info   Mean   .05  .10  .25  .50  .75  .90  .95
    188       1      24      1  23.27    16   17   19   23   26   31   32

lowest : 14 15 16 17 18, highest: 33 34 35 36 45
------------------------------------------------------------------
race
      n missing  unique
    189       0       3

White (96, 51%), Black (26, 14%), Other (67, 35%)
------------------------------------------------------------------
low
      n missing  unique
    189       0       2

No (130, 69%), Yes (59, 31%)
------------------------------------------------------------------
```

The command cut2() advantageously replaces the command cut() that is often found combined with quantile() to convert numeric variables into qualitative variables with balanced classes (and it allows the user to avoid recalling the option include.lowest=TRUE):

```
> table(cut2(birthwt$lwt, g=4))

[36.4, 50.9) [50.9, 55.5) [55.5, 64.1) [64.1,113.6]
         53           43           46           47
> table(cut2(birthwt$age, g=3, levels.mean=TRUE))

18.074 23.091 30.019
    68     66     54
```

It is also possible to specify the number of desired groups (g=) or the minimal number of observations per group (m=). The option levels.mean= makes it possible to return the class center as the factor level rather than the interval bounds.

Regarding cohort or survival data, the command event.chart() provides interesting graphical representations to visualize follow-up durations, censoring, etc. See the online help and the examples provided:

```
> example(event.chart)
```

There are also a certain number of commands enabling one to perform single or multiple imputation (see impute() and aregImpute()). A median imputation example for the data deleted in the variable ftv above is given hereafter:

```
> ftvi <- impute(birthwt$ftv)
> summary(ftvi)

5 values imputed to 0

  Min. 1st Qu.  Median    Mean 3rd Qu.    Max.
0.0000  0.0000  0.0000  0.7725  1.0000  6.0000
```

The imputed observations are indicated with an asterisk when the set of the values of the variable is displayed.

A3.2. Descriptive statistics

The command summarize() operates on the same principle as aggregate() and can provide stratified summaries for a numeric variable according to one or more qualitative variables. Unlike aggregate(), it is possible to calculate several statistics and to store them in different columns of one table of results (aggregate() stores all its results in a single column, which does not interfere with the display but does not facilitate the utilization of the results):

```
> f <- function(x) c(mean=mean(x), sd=sd(x))
> with(birthwt, summarize(bwt, race, f))
  race     bwt       sd
3 White 3102.719 727.8861
1 Black 2719.692 638.6839
2 Other 2805.284 722.1944
```

The command summary() (in fact, summary.formula()) can be used with a formula notation describing the relationship between several variables. There are three main methods (method=): "response", in which a numerical variable is summarized according to the levels of one or more variables (in the case of numeric variables, Hmisc will automatically recode the variable into four balanced classes), "cross" to calculate the conditional and the marginal means of one or several response variables according to qualitative or numeric variables (with a maximum of three) and "reverse" to summarize in a univariate manner a set of numeric (three quartiles) or qualitative (counts and proportions) variables with regard to the factor levels. The position of the variables compared to the connection operator ~ of the formula plays an important role in determining the type of summary to be performed according to the method selected. For the methods "response" and "reverse", the command plot() can be paired directly with the result returned by summary() to obtain a graphical representation of the results. A formula can also be provided and R be instructed what command has to be applied to summarize the data structure. For example, in the case low ~ race + ht, the option fun=table will automatically build two contingency tables corresponding to the crossing of the variables low with race and low with ht.

Illustrations now follow. Without specifying any option, summary() makes it possible to summarize average values of a numeric variable according to a list of explanatory numerical or categorical variables. In the case where the explanatory variable is numeric, Hmisc considers the quartiles of this variable. The frequency associated with each modality of the variables is presented in a separate column:

```
> summary(bwt ~ race + ht + lwt, data=birthwt)
Baby's weight at birth     N=189

+-------+------------+---+--------+
|       |            |N  |bwt     |
+-------+------------+---+--------+
|race   |White       | 96|3102.719|
|       |Black       | 26|2719.692|
|       |Other       | 67|2805.284|
+-------+------------+---+--------+
|ht     |No          |177|2972.232|
|       |Yes         | 12|2536.833|
+-------+------------+---+--------+
|lwt    |[36.4, 50.9)| 53|2656.340|
|       |[50.9, 55.5)| 43|3058.721|
|       |[55.5, 64.1)| 46|3074.609|
|       |[64.1,113.6]| 47|3037.957|
+-------+------------+---+--------+
|Overall|            |189|2944.587|
+-------+------------+---+--------+
```

It is possible to summarize the mean of several response variables, associated with the command cbind() and up to two explanatory variables by specifying the option method="cross". Hmisc will indicate the frequency associated with each cell, as well as the number of missing values:

```
> summary(cbind(lwt, age) ~ race + bwt, data=birthwt, method="cross")

mean by race, bwt

+-------+
|N      |
|Missing|
```

```
|lwt    |
|age    |
+-------+
+-----+-----------+-----------+-----------+-----------+--------+
| race|[ 709,2424)|[2424,3005)|[3005,3544)|[3544,4990]|   ALL  |
+-----+-----------+-----------+-----------+-----------+--------+
|White|   19      |   23      |   20      |   33      | 95     |
|     | 0         | 1         | 0         | 0         |1       |
|     | 55.55024  | 57.66798  | 62.22727  | 63.25069  |60.14354|
|     | 22.73684  | 24.78261  | 24.50000  | 24.90909  |24.35789|
+-----+-----------+-----------+-----------+-----------+--------+
|Black|    9      |    9      |    6      |    2      | 26     |
|     | 0         | 0         | 0         | 0         |0       |
|     | 65.10101  | 59.69697  | 70.83333  | 93.40909  |66.73077|
|     | 23.44444  | 20.88889  | 20.00000  | 20.50000  |21.53846|
+-----+-----------+-----------+-----------+-----------+--------+
|Other|   20      |   16      |   19      |   12      | 67     |
|     | 0         | 0         | 0         | 0         |0       |
|     | 51.22727  | 52.95455  | 58.89952  | 55.34091  |54.55224|
|     | 22.20000  | 22.68750  | 22.26316  | 22.50000  |22.38806|
+-----+-----------+-----------+-----------+-----------+--------+
|ALL  |   48      |   48      |   45      |   47      |188     |
|     | 0         | 1         | 0         | 0         |1       |
|     | 55.53977  | 56.47727  | 61.96970  | 62.51451  |59.06190|
|     | 22.64583  | 23.35417  | 22.95556  | 24.10638  |23.26596|
+-----+-----------+-----------+-----------+-----------+--------+
```

The option method="reverse" provides a summary of a set of variables, numeric or categorical, according to the levels of the factor appearing on the left of the formula. Numeric variables are summarized based on the three quartiles and the categorical variables are summarized in terms of counts and proportions. It is possible to display the mean and the standard deviation for the numeric variables by employing the command print() and by adding the option prmsd=TRUE. The number of decimals displayed is also controlled by means of the command print() (for example, round=1). Finally, the option test=TRUE associated with method="reverse" automatically adds a column, including Wilcoxon tests in the case of the comparison of two groups for the numeric variables, and Fisher's χ^2 tests in the case of categorical variables. Here is an example:

```
> summary(low ~ race + age + ui, data=birthwt, method="reverse",
          overall=TRUE)
```

Descriptive Statistics by low

```
+----------------------+---+--------------+--------------+--------------+
|                      |N  |No            |Yes           |Combined      |
|                      |   |(N=130)       |(N=59)        |(N=189)       |
+----------------------+---+--------------+--------------+--------------+
|race : White          |189|  56% (73)    |  39% (23)    |  51% (96)    |
+----------------------+---+--------------+--------------+--------------+
|    Black             |   |  12% (15)    |  19% (11)    |  14% (26)    |
+----------------------+---+--------------+--------------+--------------+
|    Other             |   |  32% (42)    |  42% (25)    |  35% (67)    |
+----------------------+---+--------------+--------------+--------------+
|Mother's age [years]  |188|19.0/23.0/28.0|19.5/22.0/25.0|19.0/23.0/26.0|
+----------------------+---+--------------+--------------+--------------+
|ui : Yes              |189|  11% ( 14)   |  24% ( 14)   |  15% ( 28)   |
+----------------------+---+--------------+--------------+--------------+
```

To obtain a multiple cross tabulation, it suffices to specify the option fun=table:

```
> summary(low ~ race + ht, data=birthwt, fun=table)
low    N=189

+-------+-----+---+---+---+
|       |     |N  |No |Yes|
+-------+-----+---+---+---+
|race   |White| 96| 73|23 |
|       |Black| 26| 15|11 |
|       |Other| 67| 42|25 |
+-------+-----+---+---+---+
|ht     |No   |177|125|52 |
|       |Yes  | 12|  5| 7 |
+-------+-----+---+---+---+
|Overall|     |189|130|59 |
+-------+-----+---+---+---+
```

All of the tables produced by Hmisc can automatically be exported in the LaTeX format and therefore reproduced in PDF version. An example of such a summary table is given in Figure A3.1. In addition, most Hmisc summary tables can also be represented in a graphical form, by making use of the command plot(), as illustrated in Figure A3.2.

A3.2.1. *Graphics commands*

The package Hmisc is essentially based on the graphics lattice, but it provides "more integrated" commands. Within the graphics commands provided by Hmisc, a distinction can be made between dot plots (Dotplot() and dotchart2()), boxplots (panel.bpplot() to be utilized with the usual command bwplot()) and scatter plots (xyplot()). We shall consider an example for the latter case only (Figure A3.3):

Description by low birth weight status (low)

		No $N = 130$	Yes $N = 59$
age	years	19.0 23.0 28.0	19.5 22.0 25.0
lwt	pounds	113.0 123.5 147.0	104.0 120.0 130.0
race : White		56% (73)	39% (23)
Black		12% (15)	19% (11)
Other		32% (42)	42% (25)
smoke : Yes		34% (44)	51% (30)
ptl : 1+		9% (12)	31% (18)
ht : Yes		4% (5)	12% (7)
ui : Yes		11% (14)	24% (14)
ftv : Yes		51% (66)	39% (23)

$a\ b\ c$ represent the three quartiles.

Figure A3.1. *Example of a statistical table generated by Hmisc*

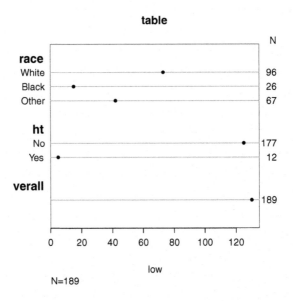

Figure A3.2. *Example of a figure generated from a table by Hmisc*

```
> f <- function(x) c(mean(x), mean(x) + c(-1, 1) * sd(x)/sqrt(length(x)))
> bwtmeans <- with(birthwt, summarize(bwt, llist(race, smoke), f))
> names(bwtmeans)[4:5] <- c("lwr", "upr")
> bwtmeans
   race smoke      bwt      lwr      upr
```

```
5 White      No 3428.750 3321.699 3535.801
6 White     Yes 2826.846 2739.970 2913.722
1 Black      No 2854.500 2699.186 3009.814
2 Black     Yes 2504.000 2302.545 2705.455
3 Other      No 2815.782 2720.133 2911.430
4 Other     Yes 2757.167 2523.327 2991.006
> xYplot(Cbind(bwt, lwr, upr) ~ numericScale(race, label="mother's
           ethnicity") | smoke,
           data=bwtmeans, type="b", ylim=c(2200, 3600),
           scales=list(x=list(at=1:3, labels=levels(birthwt$race))))
```

It should be noted that Cbind() takes a capital C to differentiate this command from the base command, and that if it exists, the label of the variable automatically replaces the name of the variable (that is the case here for the variable bwt).

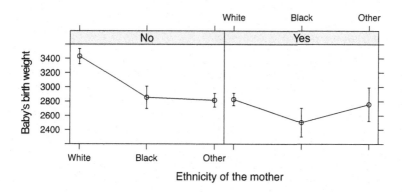

Figure A3.3. *Example of a graph of conditional means on two variables*

It is also possible to automatically annotate the different curves, the segments or the points defined by a condition variable by employing the option keys=, as in the following example (Figure A3.4):

```
> xYplot(Cbind(bwt, lwr, upr) ~ numericScale(race), groups= smoke,
           data=bwtmeans, type="l", keys="lines")
```

To change the displaying of the x-axis and to replace the numerical values by the labels of the factor, the following option may be added:

```
scales = list(x = list(at = 1:3, labels = levels(birtwht$race)))
```

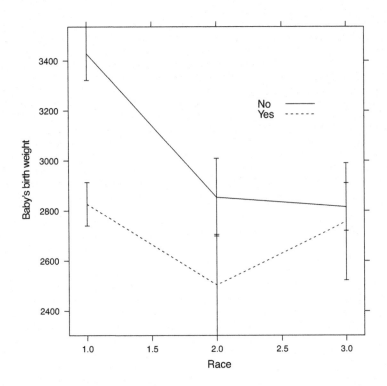

Figure A3.4. *Example of a graph of two-variable conditional means with insertion of legend*

A3.3. The essential rms commands

To use the commands in the package rms, the package must be loaded by inserting the command library(). Note that this package automatically loads the package Hmisc:

```
> library(rms)
```

For the linear regression, the command ols() must be employed. The operating principle remains the same as with the command lm(): the role of each variable is specified by using a formula (response variable ~ explanatory variables) and we indicate the data frame in which data will be searched for. In its simplest usage, results returned can be displayed by ols() with the command print() or by simply typing the name of the auxiliary variable in which the results of the model have been stored. To obtain additional information regarding the marginal effects, the residuals of the model and to be able to make use of the extensive graphics features, it is necessary to utilize the command datadist() to store the model predictors in a separate structure. Commands such as summary() or plot() will then be available.

For instance, to model the weight of infants based on the various hazard indicators of the mother, one would employ:

```
> m <- ols(bwt ~ age + lwt + race + ftv, data=birthwt, x=TRUE)
> m

Linear Regression Model

ols(formula = bwt ~ age + lwt + race + ftv, data = birthwt, x = TRUE)

Frequencies of Missing Values Due to Each Variable
 bwt   age  lwt race  ftv
  0     1    0    0    5
```

		Model Likelihood Ratio Test		Discrimination Indexes	
Obs	183	LR chi2	17.66	R2	0.092
sigma	712.1949	d.f.	5	R2 adj	0.066
d.f.	177	Pr(> chi2)	0.0034	g	254.865

```
Residuals
```

Baby's weight at birth [grammes]

Min	1Q	Median	3Q	Max
-2097.25	-440.02	40.41	490.41	1919.63

	Coef	S.E.	t	Pr(>|t|)
Intercept	2482.0826	324.5729	7.65	<0.0001
age	-0.1490	10.5951	-0.01	0.9888
lwt	10.4217	4.0481	2.57	0.0109
race=Black	-464.5878	164.4224	-2.83	0.0053
race=Other	-263.2467	119.0400	-2.21	0.0283
ftv	12.3299	51.0652	0.24	0.8095

Concerning logistic regression (including the case where the response variable has ordered modalities), the command lrm() will be used. This command selects the model according to the number of levels of the response variable: in the case where there are more than two, a proportional odds-model is used. This command operating principle is the same as in the case of ols(). In the following example, the command datadist() is illustrated and, when coupled with the rms regression models, makes it possible to calculate the effect sizes for the predictors. One can provide either a list of variables (predictors included in the model) or directly the data frame with which we are working. It is important to accompany this command by options(), as described below, to make the data available to the rms commands:

```
> ddist <- datadist(birthwt)
> options(datadist='ddist')
> m <- lrm(low ~ age + lwt + race + ftv, data=birthwt)
> print(m)
```

```
Logistic Regression Model

lrm(formula = low ~ age + lwt + race + ftv, data = birthwt)

Frequencies of Missing Values Due to Each Variable
 low  age  lwt race  ftv
   0    1    0    0    5
```

		Model Likelihood Ratio Test		Discrimination Indexes		Rank Discrim. Indexes	
Obs	183	LR chi2	12.26	R2	0.091	C	0.657
No	125	d.f.	5	g	0.680	Dxy	0.314
Yes	58	Pr(> chi2)	0.0314	gr	1.974	gamma	0.316
max |deriv|	1e-07			gp	0.136	tau-a	0.137
				Brier	0.203		

	Coef	S.E.	Wald Z	Pr(>|Z|)
Intercept	1.2883	1.0892	1.18	0.2369
age	-0.0226	0.0342	-0.66	0.5088
lwt	-0.0318	0.0146	-2.19	0.0288
race=Black	1.0074	0.5024	2.01	0.0450
race=Other	0.4831	0.3695	1.31	0.1910
ftv	-0.0396	0.1681	-0.24	0.8140

As it can be seen, the output produced by R when using lrm() instead of glm() contains additional information, notably concerning the predictive and discriminating capability of the logistic model. Here follows an application of the command summary() which will provide additional information by accessing the stored data with datadist():

```
> summary(m)
```

A likelihood ratio test can be carried out to compare a reduced model in which the factors age and ftv are excluded from the initial model by the command lrtest(). In the following instructions, a second model is built in which the variables age and ftv are removed with the command update(). This command expects an initial model, m, and a formula modified with respect to the formula of this model: in our case, the same response variable is retained and the two variables in question are removed from the right member of the formula:

```
> m2 <- update(m, . ~ . - age - ftv)
> lrtest(m2, m)
Likelihood ratio test for MLE method
Chi-squared d.f. =  , P value =
```

Finally, it is also possible to obtain a variance or deviation analysis table for the regression model through the command anova():

```
> anova(m2)
              Wald Statistics        Response: low

Factor      Chi-Square d.f.  P
lwt           5.59      1    0.0181
race          5.40      2    0.0671
TOTAL        10.13      3    0.0175
```

The command Predict() enables the automatic calculation of the values predicted by a regression model rms. Here follows an example of its usage where the objective is to calculate the log-odds values predicted by the model m2 when the mother's weight varies between 40 and 80 kg, in 5 kg increments:

```
> pm2 <- Predict(m2, lwt=seq(40, 80, by=5), race)
> head(pm2)
  lwt race      yhat      lower      upper
1  40 White -0.5338795 -1.206583  0.1388239
2  45 White -0.7013337 -1.284054 -0.1186135
3  50 White -0.8687878 -1.383708 -0.3538673
4  55 White -1.0362419 -1.515065 -0.5574193
5  60 White -1.2036960 -1.685304 -0.7220877
6  65 White -1.3711502 -1.893807 -0.8484937

Response variable (y): log odds

Limits are 0.95 confidence limits
```

As can be seen, R provides the predicted values on the log-odds scale, with a 95% confidence interval. It is, then, relatively easy to represent the predictions of this model graphically by means of xYplot():

```
> xYplot(Cbind(yhat,lower,upper) ~ lwt | race, data=pm2, layout=c(3,1),
         method="filled bands", type="l", col.fill=gray(.95))
```

For the Cox regression, this involves the command cph(). The representation of survival data always undergoes a prior conversion step with the command Surv.

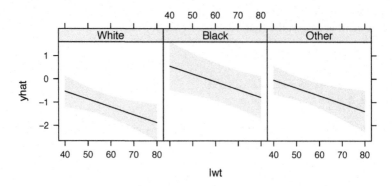

Figure A3.5. *Usage of the command xYplot()*

The bibliographic reference regarding the package rms is the book authored by Frank Harrell [HAR 01]. The following online course is constructed around this book and provides most of the ideas in a PDF document: http://biostat.mc.vanderbilt.edu/wiki/Main/CourseBios330. Ewout Steyerberg's book [STE 09] is based partly on the package rms and constitutes a good complement to the previous book. The site associated with the book is: www.clinicalpredictionmodels.org.

Bibliography

[BIL 14] BILDER C., LOUGHIN T., *Analysis of Categorical Data with R*, Chapman & Hall/CRC, Boca Raton, 2014.

[BLI 52] BLISS C., *The Statistics of Bioassay*, Academic Press, New York, 1952.

[BRE 80] BRESLOW N., DAY N., *Statistical Methods in Cancer Research: Vol. 1, The Analysis of Case-Control Studies*, IARC Scientific Publications, Lyon, 1980.

[CLE 79] CLEVELAND W., "Robust locally weighted regression and smoothing scatterplots", *Journal of the American Statistical Association*, vol. 74, pp. 829–836, 1979.

[COL 94] COLLETT D., *Modeling Survival Data in Medical Research*, Chapman & Hall/CRC, Boca Raton, 1994.

[DRA 98] DRAPER N., SMITH H., *Applied Regression Analysis*, 3rd ed., John Wiley and Sons, New York, 1998.

[DUP 09] DUPONT W., *Statistical Modeling for Biomedical Researchers*, 2nd ed., Cambridge University Press, 2009.

[EVE 01] EVERITT B., RABE-HESKETH S., *Analyzing Medical Data using S-PLUS*, Springer, New York, 2001.

[FAR 14] FARAWAY J., *Linear Models with R*, 2nd ed., Chapman & Hall/CRC, Boca Raton, 2014.

[FOX 10] FOX J., WEISBERG S., *An R Companion to Applied Regression*, Sage Publications, Los Angeles, 2nd ed., 2010.

[FOX 11] FOX J., WEISBERG S., "Cox proportional-hazards regression for survival data in R", available at www.socserv.mCaster.ca/jfox/Books/Companion/appendix/Appendix-Cox-Regression.pdf, 2011.

[FRE 63] FREIREICH E., GEHAN E., FREI E. *et al.*, "The effect of 6-mercaptopurine on the duration of steroid-induced remissions in acute leukemia: a model for evaluation of other potentially useful therapy", *Blood*, vol. 21, pp. 699–716, 1963.

[GAN 13] GANDRU C., *Reproducible Research with R and RStudio*, Chapman & Hall/CRC, Boca Raton, 2013.

[HAN 93] HAND D., DALY F., MCCONWAY K. *et al.*, *A Handbook of Small Data Sets*, Chapman & Hall/CRC, Boca Raton, 1993.

[HAR 01] HARRELL F., *Regression Modeling Strategies with Applications to Linear Models, Logistic Regression and Survival Analysis*, Springer, New York, 2001.

[HOL 73] HOLLANDER M., WOLFE D., *Nonparametric Statistical Inference*, John Wiley and Sons, New York, 1973.

[HOS 89] HOSMER D., LEMESHOW S., *Applied Logistic Regression*, John Wiley and Sons, New York, 1989.

[HOT 09] HOTHORN T., EVERITT B., *A Handbook of Statistical Analyses Using R*, 2nd ed., Chapman & Hall/CRC, Boca Raton, 2009.

[IHA 96] IHAKA R., GENTLEMAN R., "R: A language for data analysis and graphics", *Journal of Computational and Graphical Statistics*, vol. 5, no. 3, pp. 299–314, 1996.

[LOP 94] LOPRINZI C., LAURIE J., WIEAND H. *et al.*, "Prospective evaluation of prognostic variables from patient-completed questionnaires", *Journal of Clinical Oncology*, vol. 12, no. 3, pp. 601–607, 1994.

[MUR 05] MURRELL P., *R Graphics*, Chapman & Hall/CRC, Boca Raton, 2005.

[PEA 05] PEAT J., BARTON B., *Medical Statistics: A Guide to Data Analysis and Critical Appraisal*, 2nd ed., John Wiley and Sons, New York, 2005.

[PEP 80] PEPE M., *The Statistical Evaluation of Medical Tests for Classification*, Oxford University Press, New York, 1980.

[RAB 04] RABE-HESKETH S., EVERITT B., *A Handbook of Statistical Analyses using Stata*, 3rd ed., Chapman & Hall/CRC, Boca Raton, 2004.

[SAR 08] SARKAR D., *Lattice: Multivariate Data Visualization with R*, Springer, New York, 2008.

[SCO 08] SCOTT I., GREENBURG P., POOLE P., "Cautionary tales in the clinical interpretation of studies of diagnostic tests", *Internal Medicine Journal*, vol. 38, pp. 120–129, 2008.

[SEL 98] SELVIN S., *Modern Applied Biostatistical Methods using S-PLUS*, Oxford University Press, New York, 1998.

[SHA 12] SHAHBABA B., *Biostatistics with R: An Introduction to Statistics through Biological Data*, Springer, New York, 2012.

[SPE 08] SPECTOR P., *Data Manipulation with R*, Springer, New York, 2008.

[STE 09] STEYERBERG E., *Clinical Prediction Models: A Practical Approach to Development, Validation and Updating*, Springer, New York, 2009.

[STU 08] STUDENT, "The probable error of a mean", *Biometrika*, vol. 6, no. 1, pp. 1–25, 1908.

[THE 99] THERNEAU T., "A package for survival analysis in S", available at https://cran.r-project.org/web/packages/survival/citation.html, 1999.

[VEN 02] VENABLES W., RIPLEY B., *Modern Applied Statistics with S*, 4th ed., Springer, New York, 2002.

[VER 11] VERZANI J., *Getting started with RStudio*, O'Reilly Media, Sebastopol, 2011.

[VIT 05] VITTINGHOFF E., GLIDDEN D., SHIBOSKI S. *et al.*, *Regression Methods in Biostatistics. Linear, Logistic, Survival and Repeated Measures Models*, Springer, New York, 2005.

[WIC 11] WICKHAM H., "The split-apply-combine strategy for data analysis", *Journal of Statistical Software*, vol. 40, no. 1, 2011.

[WIC 14] WICKHAM H., "Tidy data", *Journal of Statistical Software*, vol. 59, no. 1, 2014.

[WIL 73] WILKINSON G., ROGERS C., "Symbolic description of factorial models for analysis of variance", *Applied Statistics*, vol. 22, pp. 392–399, 1973.

[WIL 05] WILKINSON L., *The Grammar of Graphics*, Springer, New York, 2005.

Index

Printed in the United States
By Bookmasters